Threats to Humanity

by

Leroy W. Dubeck, Ph.D.
and
Suzanne E. Moshier, Ph.D.

Threats to Humanity

by Leroy W. Dubeck, Ph.D.
and
Suzanne E. Moshier, Ph.D.

Copyright © 2008 by
Leroy W. Dubeck, Ph.D.
and Suzanne E. Moshier, Ph.D.

First Edition, Printed September, 2008
by Ishi Press International in New York

All rights reserved according to International Law. No part of this book may be reproduced by any mechanical, photographic or electronic process nor may it be stored in a retrieval system, transmitted or otherwise copied for public or private use without the written permission of the authors.

ISBN 0-923891-56-0
978-0-923891-56-5

Ishi Press International
1664 Davidson Avenue, Suite 1B
Bronx NY 10453
USA
917-507-7226

Printed in the United States of America

TABLE OF CONTENTS

Introduction   Page 5

Chapter 1: Sixty-Five Million Years and Counting: The Science of *Deep Impact* and *Armageddon*   Page 7

Chapter 2: Nuclear Terrorism: The Science of *The Peacemaker* and *The Sum of All Fears*   Page 31

Chapter 3: Global Warming and the Greenhouse Effect: The Science of *The Day After Tomorrow* and *An Inconvenient Truth*   Page 53

Chapter 4: Artificial Intelligence: The Science of *I Robot* and *Blade Runner*   Page 77

Chapter 5: Volcanoes and the Science of *Dante's Peak* and *Volcano*   Page 101

Chapter 6: Pandemics and Modern Plagues: The Science of *Outbreak* and *The Andromeda Strain*   Page 123

ABOUT THE AUTHORS

Leroy W. Dubeck, Ph.D., is a Professor of Physics at Temple University in Philadelphia, Pennsylvania. He has taught a Temple University Course, Science and Science Fiction in Film, more than 50 times, both onsite and online. He has been Principle Investigator on a number of National Science Foundation grants to develop the use of science fiction films as a teaching tool at both the college and precollege level. He has been co-author of many textbooks and articles describing the use of science fiction films to teach science.

Suzanne E. Moshier, Ph.D., is a Professor of Biology at the University of Nebraska at Omaha in Omaha, Nebraska. She has co-authored many textbooks and articles with Dr. Dubeck on the use of science fiction films to teach science, participated in National Science Foundation funded research and teacher development using this technique, and has taught Chautauqua and other workshops devoted to it.

The writing of this textbook was supported by a study leave from Temple University.

Cover Image: Barringer Crater, Courtesy: D. Roddy (LPI).

INTRODUCTION

In this book we will examine six threats to humanity. Each of these threats can have an impact not just on one country but on all humanity should it materialize. For some of the threats such as an impact by an asteroid or meteor or global warming the worldwide impact is obvious. For others, such as a pandemic or nuclear terrorism it may not be quite as obvious but it is none the less real. For a volcanic eruption to have a worldwide impact may seem unlikely to most readers, but as that chapter indicates, it has happened in the past. Finally artificial intelligence is not likely to be developed in only one country. It too will be worldwide.

While we hope that a global catastrophe does not occur, the best hope for humanity to prevent or survive such a catastrophe is to be aware of the danger and to do all in our power to prevent it. When it comes to these threats, ignorance is definitely not bliss.

Each chapter begins with a discussion of the scientific principles involved in understanding the threat and therefore how to prevent it from happening or mitigating its consequences if it does occur. Then we review two films which depict the threat. The plot summary is first presented, usually followed by movie trivia about the film.

The next section, entitled SciFi Science vs. Real Science, discusses the way aspects of the threat are depicted in the two films and compares these portrayals with scientific reality.

Finally each chapter concludes with a number of questions that the reader should consider. They may be part of a course's assigned workload if this textbook is adopted for that course.

The information that we have provided about these topics is "Internet friendly" in the sense that the reader should be able to search the Internet using key words and technical terms that appear in the text in order to get more detailed information about many of the topics included in this book.

We have enjoyed screening many films to select the films included in this book. We hope that you will enjoy watching them (again if you have seen them before) and comparing them with real science. We also hope that studying these films will help you to critically analyze other "information" from the mass media to determine whether that information is valid. Using the approach in this book can provide a platform for a better understanding of the world around us. Remember, knowledge is power.

SIXTY-FIVE MILLION YEARS AND COUNTING

THE SCIENCE OF *DEEP IMPACT* AND *ARMAGEDDON*

About sixty five million years ago the dinosaurs were the dominant species on Earth. Virtually overnight, suddenly they and untold other species disappeared from the planet. Scientists believe that Earth was hit by a comet or asteroid about 5 or 6 miles in diameter. Other such extinction level events have occurred in the past (some were likely due to other collisions with comets or asteroids) and will occur again in the future. All that is unknown is when, but we may be overdue for the "big one."

ASTEROIDS AND COMETS

A <u>meteoroid</u> is a small solid object in outer space, usually coming from an asteroid or comet. Rarely, some meteoroids are rocks blasted off Mars or the Moon by a collision with an asteroid.

A <u>meteor</u> is a meteoroid that has entered Earth's atmosphere. Often these are called "shooting stars."

A <u>meteorite</u> is a meteoroid that has survived its passage through the Earth's atmosphere and has reached the surface of the Earth.

TWO KINDS OF METEOROIDS: COMETS AND ASTEROIDS

<u>Comets</u> are frozen icebergs which consist of a combination of

Chapter One: Sixty-Five Million Years and Counting

frozen gases, such as ammonia and methane, with ice and dust or more solid materials embedded in the frozen gases.

Asteroids are more solid objects consisting of either stone, iron or a combination of stone and iron.

Comets are easier to see in the sky because as they approach the Sun they start evaporating. Most originate from a region called the Oort Cloud. It exists far beyond the orbit of Pluto. When a comet is far from the Sun it consists of only a nucleus, i.e. a hard solid object anywhere from one mile in diameter to dozens of miles wide. It has no head or tail. Since the surface of a comet's nucleus is dark, it would be difficult to see from the Earth if it remained only a hard solid object.

Figure 1. A comet near the sun
Comet Kohoutek. Courtesy NASA

Chapter One: Sixty-Five Million Years and Counting

However, as a comet approaches the Sun, its heat evaporates some of the frozen gases and dirt which then forms a kind of cloud around the nucleus that is called the <u>coma</u> of the comet. The nucleus and the coma are sometimes referred to as the <u>head</u> of the comet. The pressure of sunlight pushes the dust particles in a direction opposite to the sun, producing the comet's dust tail. A comet can also have a <u>plasma tail.</u>

The plasma tail is produced by the ionization of gas atoms when struck by sunlight. <u>Ionized</u> <u>atoms</u> have a net electric charge on them. An electrically neutral atom has as many positively charged particles, called <u>protons,</u> in its nucleus or center as there are negatively charged particles, called <u>electrons,</u> in orbits about the nucleus. An excess or deficiency of electrons makes the atom have a net electrical charge. Ionized gases are subject to the pressure of the <u>solar wind,</u> an invisible stream of charged particles emitted by the sun. The solar wind pushes the cometary gas away from the sun, forming the plasma tail. Observers can easily see the dust tail or plasma tail which may extend for millions of miles behind the head of the comet.

Asteroids and comets are of concern to us only if they pass close to the Earth so that a small deviation in their path (caused perhaps by a collision with another asteroid or comet) might send it hurtling towards us.

Earth has had many collisions in the recent past. Scientists believe that the frequency of collisions is greater for smaller objects and less frequent for larger ones. Everyday the Earth's atmosphere is struck by countless small meteoroids which burn up harmlessly in the atmosphere. Sometimes the meteoroid hits the ground producing a crater. One estimate is that we are struck by a fairly large object about once each ten thousand years. Both comets and asteroids have <u>kinetic energy,</u> which is the energy any moving object has. Before the object comes to rest, it must give up this energy, which is sometimes converted into heat or it is used to smash the things which it hits to

Chapter One: Sixty-Five Million Years and Counting

come to rest, as, for example, a moving car smashing into another car stopped at a red light.

A comet or asteroid has much greater kinetic energy per pound than any car because the asteroid or comet may be moving in excess of 20,000 miles per hour when it slams into the Earth. Cars rarely go faster than 100 miles per hour. One pound of a comet or asteroid moving at 20,000 miles per hour has 40,000 times the kinetic energy of one pound of a car going at 100 miles per hour.

EXAMPLE OF PAST COLLISIONS

Figure 2: Barringer Crater Credit D. Roddy (LPI)

Chapter One: Sixty-Five Million Years and Counting

Barringer Crater is near Winslow, Arizona and is 500 feet deep and 4,000 feet wide. It was created by a meteorite weighing at least 30,000 tons. It struck the Earth between 25,000 and 50,000 years ago. The tremendous kinetic energy it possessed when it struck was equivalent to the explosion of 20 to 40 million tons of TNT, more than 1,000 times greater than the blasts that destroyed Hiroshima and Nagasaki at the end of World War II. The air blast killed or badly injured all animals out to a distance of at least 12 miles from the impact point.

The Tunguska Explosion occurred even more recently, in 1908, in a forested region near the Tunguska River in Siberia. The explosion occurred several kilometers above the surface, similar to a mid-air nuclear blast, and was powerful enough to level trees in an area of about 800 square miles. One estimate of the force of the explosion was that it was equivalent to 10-20 megatons of TNT exploding. A megaton is one million tons. Since the collision did not leave a large impact crater it probably was caused by a comet exploding above the surface. Had this explosion occurred over a population center in Europe, it could have killed millions of people.

The Comet Explosion over North America 11,000 years ago is a relatively recent possible explanation for the actual extinction of many mammals such as the saber toothed tiger and the woolly mammoths, as well as early human settlers called the Clovis civilization. An educational television program speculated that a large comet exploded in the air above North America (hence no crater) and scattered burning remnants for thousands of miles igniting a continent wide fire storm. As evidence to support this hypothesis, scientists reported in the program finding a layer of ash (from burning trees, brush etc) at Clovis settlements across North America, all ash deposited at the same time. If this explanation is correct, it is a stark example of what such a celestial impact could to the human race.

11

Chapter One: Sixty-Five Million Years and Counting

WHEN WILL THE NEXT MAJOR COLLISION OCCUR?

No one can predict when Earth will next be struck by a large comet or asteroid.

Little Known Fact: During December, 2004 while the world was riveted by the Indian Ocean Tsunami that killed more than 250,000 people, another rare but even more devastating event was developing virtually unnoticed. In June 2004, an asteroid designated 2004MN4 was reported by observers using the Bok telescope at Kitt Peak, Arizona. By December 23, 2004 it appeared that while the asteroid would not likely hit Earth in the near future, there was a 1 in 300 chance of an impact on Friday April 13, 2029. By Christmas day the probability of impact had risen to 1 in 47! Then it rose further to 1 in 37, about the same probability as rolling double 6's in dice. Astronomers felt relief as they discovered the asteroid's faint presence on plates taken of the sky in March 2004. These plates, when integrated with the more recent data, showed that it would definitely miss the Earth on April 13, 2029. It will pass the Earth at less than one tenth the distance of the Moon. Depending on its exact orbit past the Earth on April 13, 2029, the asteroid and the Earth could come into collision exactly 7 years later, on April 13, 2036.

How large is 2004MN4? It is approximately 320 meters (about 1100 feet) in diameter. It would hit with the explosive force of nearly a thousand megatons of TNT, or about 50,000 Hiroshima type atom bombs and could easily kill tens of millions of people.

Because the collision with a large comet or asteroid would be more devastating than any other natural disaster, it is a natural topic for disaster films. Among the earlier films were *When Worlds Collide* and *Meteor*. In this chapter we compare the presentation in two more recent films, *Deep Impact* and *Armageddon*, both released in 1998.

Chapter One: Sixty-Five Million Years and Counting

FILM SUMMARY

DEEP IMPACT

Paramount Pictures and DreamWorks Pictures (US), 1998, color, 121 minutes

Cast: Robert Duvall (Captain Spurgeon Tanner), Tea Leoni (Jenny Lerner), Vanessa Redgrave (Robin Lerner), Maximillian Schell (Jason Lerner), Elijah Wood (Leo Biederman) and Morgan Freeman (President Tom Beck)

Director: Mimi Leder

The film begins with a high school astronomy student, Leo Biederman, taking a photograph of an unidentified object using his telescope. The photograph is sent to a scientist at a nearby observatory who panics when his computer predicts a collision with Earth. He is killed in a traffic accident while rushing to take the information to another location.

The film then moves forward one year and follows a news reporter, Jenny Lerner, who stumbles on a story about "Ellie," which she at first thinks is the name of the mistress of a high government official. Soon she and the world realize it stands for E.L.E. or extinction level event. There is a 7 mile long comet on a collision course with Earth and it will strike in about one year.

Chapter One: Sixty-Five Million Years and Counting

Figure 3: *Deep Impact*. Two astronauts placing nuclear explosives under the surface of a comet on a collision course with Earth. Courtesy of Paramount Pictures/Photofest

The President announces the building of a spaceship, the Messiah, which will travel to the comet and place nuclear bombs beneath its surface that will deflect it from the Earth when they detonate. The Messiah is propelled by the Orion propulsion system, which uses a series of small nuclear explosions which push against a massive plate at the rear of the spaceship. The plate is attached to the crew's quarters through a shock absorbing system to reduce the impact of these explosions on the crew and also to shield them from radiation emitted by the nuclear explosions.

The pilot that will land the crew and the nuclear bombs on the surface of the comet is Spurgeon Tanner. The crew plants four bombs each with the explosive power of 5 megatons of TNT but are not able

Chapter One: Sixty-Five Million Years and Counting

to fully complete the task of boring each hole 100 meters (about 300 feet) into the comet's surface and placing each bomb into the bore hole. This needs to be done before the rotation of the comet exposes the surface they are working on to sunlight which causes explosive outgassing from the comet's surface. One crew member is killed and another injured but the bombs are detonated. Unfortunately they merely break the comet in two pieces, one is 1.5 miles in diameter and the second about 5 miles in diameter.

Next the US and Russia attempt to deflect/destroy the comets by firing ICBM's (Inter-continental Ballistic Missiles) at them from the Earth's surface. The missiles miss. Then all that is left is that a million Americans (most selected by a lottery) will be evacuated to a vast underground "ark" which hopefully contains enough food to survive the post impact "winter." The remainder of the public is doomed.

Meanwhile the Messiah astronauts, now enroute back to Earth, see a fissure in the larger comet. They dive their spaceship into the fissure and detonate a second set of nuclear warheads. This explosion shatters the comet into a million pieces which burn up harmlessly in the atmosphere. The smaller of the comets hits the North Atlantic and creates a gigantic tsunami which is thousands of feet high at the coastline and which travels 600 miles inland. The film switches back and forth between the Messiah, the White House and two families, the Biedermans and the Lerners as they prepare for the unthinkable.

MOVIE TRIVIA

DEEP IMPACT

Mimi Leder, the Director of *Deep Impact,* prepared for the shooting of the film by viewing *On the Beach*, the 1959 film about the end of the world after a nuclear holocaust spreads lethal radioactive

Chapter One: Sixty-Five Million Years and Counting

fallout worldwide. She said that she wanted to get a feeling as to how humans would react to such an impending catastrophe. One of the characters in *Deep Impact*, Jenny Lerner's divorced mother, does kill herself as the impending doomsday scenario approaches. As it turns out she might have survived the impact of the smaller comet had she been able to get far enough inland (she lived in Washington DC which was destroyed by the tsunami).

Deep Impact also deals with teenagers facing the catastrophe: A very young Leo Biederman marries his 14 year old classmate since he has been preselected for admittance to the Ark and this will entitle his wife to be saved. Finally the heroes of the film are of course the astronauts who sacrifice themselves by diving their spaceship into the second comet to destroy it with nuclear bombs.

Elijah Wood was only 16 when the film was made. He needed to first get a driver's license so that he could do the scene on the motorcycle in which he rescues his "wife" from the tsunami!

ARMAGEDDON

Touchstone/Disney (US), 1998, color, 150 minutes

Cast: Bruce Willis (Harry Stamper), Ben Affleck (A.J. Frost), Billy Bob Thornton (Dan Truman) and Liv Tyler (Grace Stamper)

Director: Michael Bay

The film commences with the oil drilling crew of Harry Stamper, his number one assistant, A.J. Frost, his daughter Grace and others on an oil rig in the ocean. Government officials land to say it is vital that they come to NASA. There they learn that an asteroid "the size of Texas" is on a collision course with Earth and will impact in 18 days. The drillers are asked to instruct a group of astronauts on how to drill an 800 foot deep hole into the surface of the asteroid and plant a

Chapter One: Sixty-Five Million Years and Counting

nuclear bomb in it in order to split the asteroid in two and send each half on a new orbit that misses the earth. Stamper says that the time available is too short to train the astronauts as drillers so he and his team instead agree to go aboard two spaceships that will attempt to plant the nuclear bombs. In the story only one bomb is needed to split an asteroid the size of Texas in two, and sending two ships gives NASA twice the chance of success.

The film depicts the training of the new "astronauts." The spaceships are launched and travel around the Moon to gain speed and then approach the comet from behind. One of the spaceships (containing AJ Frost and his team) crashes on the asteroid, killing some of the crew. The survivors try to reach the landing site of the other ship by using a rover, called an Armadillo. They arrive in time to help complete the bore hole. However the remote detonator malfunctions and so Stamper stays behind and detonates the nuclear bomb by hand. It splits the asteroid into two pieces both of which miss Earth by 400 miles.

MOVIE TRIVIA

Michael Bay, Director of *Armageddon*, had a number of insightful observations about the audience for his film. He noted that the film was mainly entertainment, not a science documentary. Reportedly he said that most of the audience would never believe that a comet or asteroid only 5 or 6 miles across could wipe out humanity – the object had to be bigger. Thus the asteroid in the film is much bigger – it is the size of Texas. Similarly, although his technical advisers said that it would take 10,000 hydrogen bombs to split such an object, just one does the job in the film, a reflection perhaps of the lack of understanding by the general theater audience about what a hydrogen bomb could and could not do. You, the reader, now know the answer!

Bruce Willis reportedly asked one question of the Director

Chapter One: Sixty-Five Million Years and Counting

before agreeing to play the role of Harry S. Stumper. Would the Director guarantee that his character dies at the end of the film (saving the whole world of course)?

Although NASA offered to let *Armageddon* film at its actual control center, the buildings looked more like a 1950's community college than a state of the art facility. Thus the film crew put a large NASA sign on the outside and inside lobby of a Chinese herb factory and, voila, one had a much newer looking NASA headquarters.

Similarly the *Armageddon* crew built its own version of the NASA control center to look more modern.

SCI FI SCIENCE vs. REAL SCIENCE

COMET/ASTEROID SIZE AND DISCOVERY

DEEP IMPACT. 7 miles across, discovered 2 years before impact

ARMAGEDDON. The size of Texas, at least 600 miles in diameter, discovered 18 days before impact.

Real science. The size of a comet or asteroid can vary. However, a 600 miles wide object is the size of Ceres, the largest asteroid we have found to date. It is in an orbit between Mars and Jupiter and was first seen by astronomers centuries ago. It is very unlikely that any object that size would not be seen decades before a collision with the Earth.

The 7 mile long comet in *Deep Impact*, weighing about 500 billion tons, might not be seen until two years before impact, however. By comparison astronomers have reported a .8 mile diameter asteroid which has a one in 300 chance of striking the Earth, but not until March 16, 2880! Knowing centuries in advance of a possible collision

Chapter One: Sixty-Five Million Years and Counting

would give us ample opportunity to change the future path of an asteroid or comet.

METHOD OF DEFLECTING/DESTROYING THE ASTEROID OR COMET

DEEP IMPACT. Detonate four hydrogen bombs, each with the explosive power of 5 million tons of TNT, 300 feet below the surface, to deflect/destroy the comet

ARMAGEDDON. Detonate one hydrogen bomb (explosive power unstated), 800 feet below the surface, which will split the asteroid into two pieces, each of which will miss the Earth.

Real science. Detonating hydrogen bombs on the surface or below the surface are two possible solutions to avoiding a global killer. Exploding bomb(s) on the surface would give the asteroid or comet a recoil in the direction opposite to the blast, in the same way that a rifle recoils when a bullet is fired from its muzzle. This kind of a push might be enough to change the future path of the asteroid or comet if it takes place long before its anticipated collision with the Earth.

ARMAGEDDON. The explosion was detonated after the asteroid had passed the Moon (which is about 250,000 miles from Earth). Since the asteroid was moving on a collision course with Earth at a reported 22,000 miles per hour, and the landings of the spaceships only take place after the asteroid had passed the Moon, followed by several hours of drilling, the asteroid must have moved to within 100,000 miles of Earth. This is too close for an object the size of Texas to be sufficiently deflected by one hydrogen bomb exploding a few hundred feet below its surface.

DEEP IMPACT. It is not clear as to whether the bombs are expected to shatter the comet or merely alter its course. It is several months away from impact when the bombs are detonated and it is

19

Chapter One: Sixty-Five Million Years and Counting

more realistic that four hydrogen bombs could shatter or deflect an object 7 miles long.

Other possible ways of deflecting an incoming comet or asteroid include depositing a "mass driver" on the surface of the comet or asteroid that would hurl part of the object into space causing the object to recoil in the opposite direction and hence change its orbit. Also one could plant large rockets on the surface of the comet or asteroid. Their ignition would do the same thing as the mass driver, but in this case hurl gases at high speed in one direction so that the comet or asteroid recoils, changing its future orbit. If one had enough time before impact one could even paint the surface of the object white in order to change its heat radiation, thus changing the way it reflects sunlight. Over a long enough period of time this would also change its orbit. Also one could attach a large metallic sail to the object and have sunlight push it into a new orbit.

POTENTIAL DAMAGE TO BE DONE BY THE IMPACT

DEEP IMPACT. The collision with the 7 mile long comet would release energy equivalent to the explosion of about 100 million megatons or more than 10,000 tons of TNT for each man woman and child on the planet. As the film stated the impact would kill many by blast (or tsunami) but the dust and debris hurled into the atmosphere would block out all sunlight resulting in the eventual death of all plants and animals on the planet in matter of a few months.

ARMAGEDDON. The impact with an asteroid the size of Texas would end all human and even microbial life on the planet.

Real Science. Both films are correct in their prediction of the destruction if collision occurs. The object which is believed to have caused the destruction of the dinosaurs and many other animals world wide 65 million years ago was 5 or 6 miles in diameter. It struck off the coast of the Yucatan peninsula in Mexico producing a crater about

Chapter One: Sixty-Five Million Years and Counting

120 miles across.

TRANSPORATION TO THE COMET/ASTEROID

DEEP IMPACT. The crew is lifted by the space shuttle to the new spaceship, the Messiah. It is constructed in about 14 months after the discovery of the comet. It is powered by small nuclear explosions on a massive plate at the rear of the ship that attaches to the crews quarters through a shock absorbing system.

ARMAGEDDON. Two NASA shuttles carry the two drilling crews to the Russian space station where their space shuttles are to receive more liquid fuel. Then the shuttles proceed around the Moon and onto the asteroid's surface.

Real Science: Scientists have speculated about using the Orion type spaceship to reach nearby stars! However, we do not at present have the technology to build the controlled nuclear detonation part of the ship. Building such a spaceship, in orbit, within 14 months after discovering the comet also seems rather short, but if unlimited funding was available, it might be achieved providing that the controlled nuclear detonation issue was resolved.

SURFACE OF THE COMET OR ASTEROID

DEEP IMPACT. The comet was surrounded by debris and the surface had many fissures that were outgassing when the sunlight hit the surface. The comet was rotating every 7 hours. Outgassing of gases from the surface of the comet as sunlight reaches it is vividly displayed.

ARMAGEDDON. The surface of the asteroid is pictured as having many sharp rock formations on it. The site where the shuttle

Chapter One: Sixty-Five Million Years and Counting

lands is mainly metal, which chews the drill bits.

Real Science. Both films are reasonably accurate here with one exception. The surfaces are seen to be reasonably bright, reflecting the sunlight hitting them. Actually comets have a very dark surface. However, depicting the surfaces' light reflecting properties accurately would have made it impossible to see what was going on in the films. Since these are films, they both had to change the reflectivity properties of the surface.

THE GRAVITY OF THE COMET OR ASTEROID

DEEP IMPACT. The comet is depicted as having very little gravity. However, the astronauts still move about the surface reasonably normally, whereas they would be propelled away from the comet by simply pressing down hard on the surface as they walk. When one astronaut is hit by a stream of gases emitted from the surface when the sunlight strikes it, he is propelled into outer space as he would be on an actual comet which had nearly zero gravity. Also the astronauts are accurately depicted as being weightless during their trip to and from the comet.

ARMAGEDDON. The gravity on the surface of the asteroid is stated as being small so the astronauts have small jets on the top of their suits which propel air upward. The recoil pushes the astronaut down onto the asteroid's surface.

Real Science. Both films are reasonably accurate in portraying the nearly gravitationless comet and the huge asteroid which would have a surface gravity only a few percent of that at Earth's surface. The gravity at the surface of any planet, comet or asteroid is determined by two factors: the total amount of matter in the object and how large is the object. An object of the same size made of denser materials, such as heavy metals, will have a greater gravity at its surface than an object made of less dense materials, such as a comet

Chapter One: Sixty-Five Million Years and Counting

consisting in part of frozen gases. <u>Density</u> is the amount of matter per unit volume. Thus the gravity at the surface of the Moon is only 16% of that at the surface of the Earth, both because the Moon is smaller than the Earth (with a diameter of 2160 miles while the Earth is about 7900 miles in diameter) and is also on average less dense than the Earth.

Armageddon also shows the giant pool which NASA uses to try to simulate zero gravity when an astronaut is submerged. This gives the astronauts the experience of working in an environment approaching zero gravity.

Interesting Point: In shooting the film, Bruce Willis and Ben Affleck were filmed in the actual pool, the first ever non NASA personnel that were allowed to use that facility.

RUSSIAN SPACE STATION

ARMAGEDDON. The Russian Station is pictured as firing its rockets to start rotating. The Rotation will create the same sensations as gravity, allowing the astronauts to work faster in refueling. However docking to a rotating space station would be quite a feat.

The two space shuttles stop at the Russian space station to refuel. According to the film this involves taking on liquid oxygen. Due to the dilapidated equipment and the Russian cosmonaut, who seems a little crazy, an accident occurs which causes a fire and destroys the space station.

Real Science. Rotating a space station will create the same sensation as gravity. This has been long known by scientists and depicted in many prior films. For example, in the film 2001 the huge circular space station is rotated at just the right speed to create an artificial gravity equivalent to that on the surface of the Earth.

Chapter One: Sixty-Five Million Years and Counting

The fuel is depicted as coming from one source, liquid oxygen. Actually two liquid components, oxygen and hydrogen would need to be taken on the space shuttles and then brought together in the shuttles' rockets to provide thrust for the remainder of their flight.

In outer space, which is where the space station is located, there is no air, hence no oxygen, and an object cannot burn without oxygen. Thus the fire seen burning throughout the space station as it explodes would not have looked that way in real life.

EFFECT OF THE BOMBS DETONATING

DEEP IMPACT. There are two sets of 4 hydrogen bombs each that are detonated. Each hydrogen bomb used has the explosive power equivalent to that of detonating 5 megatons of TNT.

The explosion of the first set of hydrogen bombs, each buried some 300 feet below the surface, splits the 7 mile comet into 2 pieces. The second set of 4 hydrogen bombs detonated in a 2 mile deep fissure inside the 5 mile diameter second comet shatters it into a million pieces that burn up harmlessly in the atmosphere.

ARMAGEDDON. One hydrogen bomb inserted 800 feet into the asteroid splits the Texas sized asteroid into 2 pieces and gives enough of a sideways push to each so both pieces miss the Earth.

Real Science. The first set of nuclear bombs could indeed split a 7 mile long comet into two pieces as depicted in Deep Impact. The comet is much less dense than an asteroid. The explosion of the second set of bombs inside of the larger of the two comets created by the first set of bombs is very unlikely to shatter the comet into so many small pieces that none are large enough to actually strike the Earth. The explosion occurs only hours before the second comet is expected to hit the Earth. While some parts of it would have been shattered into pieces small enough to burn up in the atmosphere and some pieces may have

Chapter One: Sixty-Five Million Years and Counting

had their trajectory changed enough to miss the Earth, many fragments would still have struck the Earth causing tremendous destruction, although perhaps not being an extinction level event.

The breaking in two of the Texas sized asteroid in Armageddon by one hydrogen bomb (which is only several feet long and narrow enough that it can be inserted into a relatively small bore hole) is pure fantasy. The bomb would make a crater which is very small compared to the size of the asteroid. It also would not deflect the asteroid from hitting the Earth because any change in its orbit would be too small to prevent a collision with the Earth, killing everyone.

An analogy to what might happen is to consider the following. Imagine a car traveling at 100 miles per hour heading straight towards a 100 foot wide building. When the car is about 1000 feet from the building someone detonates a firecracker on the roof of the car. The firecracker will make a loud noise and may even make a dent in the roof but the car will remain in one piece and will still hit the building head on less than 10 seconds after the fire cracker is detonated.

In order to split the Texas sized asteroid in two, one scientist estimated that it would take 10,000 hydrogen bombs all buried in a straight line of bore holes across the length of the asteroid and all exploding simultaneously.

There is a single bomb that in theory might have the explosive power to split or deflect the asteroid. This bomb would consist of matter and anti-matter that are brought into contact at detonation. Here the matter-anti matter annihilates each other and the mass is completely transformed into energy according to Albert Einstein's famous formula, $E=mc^2$ where E is the energy equal to a given amount of mass m and c is the speed of light in a vacuum. A hydrogen bomb converts a small fraction of its mass into energy according to the Einstein formula. It is this small fraction of matter being converted into pure energy that provides for the enormous explosive power of a hydrogen bomb. The matter-anti-matter bomb would be perhaps 10,000 more powerful than a hydrogen bomb of the same size. However, we cannot build such a device at the present time because the anti-matter immediately explosively interacts with any "normal" matter it contacts. So this bomb does not exist, nor does the film refer to such a device as being the bomb used to split the asteroid.

HITTING OF THE ADVANCE PIECES OF ASTEROID

ARMAGEDDON. The pieces preceding the asteroid strike New York City causing a huge amount of destruction. The NASA head says that they are only small fragments the size of a basketball or a car. Another fragment hits the ocean near Shanghai, causing a massive amount of destruction.

Real Science. The size of a small impactor striking the Earth will be much less than the size of the object that enters the atmosphere since most of a "small" object will burn up in the atmosphere. One study suggested that impactors of less than about 30 feet in diameter would probably cause no damage because most of them will burn up or break up in the atmosphere. Also the direction of the incoming object is important. If the object enters the atmosphere at a low angle relative to Earth's surface, it will travel further in Earth's atmosphere (losing most of its speed and hence its kinetic energy, before it impacts the surface. The impact will then be much less destructive than that of an object that dives vertically through the atmosphere to the Earth's surface.

WATER IMPACT OF THE SMALLER COMET

DEEP IMPACT. The high point in the special effects of *Deep Impact* is when the smaller comet strikes the Atlantic Ocean producing a wave several hundred feet high that moves at a speed of 1100 miles per hour through the water. As it reaches the shoreline the wave height grows to thousands of feet. The wave continues inland for up to 600

Chapter One: Sixty-Five Million Years and Counting

miles.

Real Science. An ocean impact by a comet of diameter 1.5 miles or so will cause a monster tsunami. As the wave nears the shore, depending upon the depth of the ocean floor near the shoreline, the wave will grow in height as depicted in the film. How far the wave will travel inland will depend upon the terrain (mountainous or not and the amount of friction the terrain causes on the wave). One scientist estimated that along the Atlantic coast the wave would only travel a tenth of the 600 miles inland reported in the film.

HOW THE GOVERNMENT PRESENTS THE NEWS TO THE PUBLIC

ARMAGEDDON. Since the government only learns of the impact 18 days before collision, it does not need to do much to keep the information from the public beyond restricting the release of the observations by astronomers. In the film it is stated that only 8 telescopes in the world can see the incoming asteroid and the government operates 7 of them. The news is released to the public as the astronauts take off to intercept the asteroid, but this is only hours before impact so maintaining order is not that crucial at that point in time.

DEEP IMPACT. The comet is identified about 2 years before impact. The government keeps knowledge of it a secret for about one year during which most of the work on building the spaceship, the Messiah, is completed in secret. When the news is announced it is done in the most optimistic way with the President referring to the crew of the Messiah as the people who will stop the comet.

When the first set of nuclear bombs only splits the comet in two, both of which are still headed for a collision with Earth, the President says that Titan missiles will be fired at the comet by the

27

Chapter One: Sixty-Five Million Years and Counting

United States and Russia to deflect or destroy them. This can only be done within a day or so before impact. The President tries to convey optimism that the missiles will stop the comets. At the same time he announces the building of an underground ark that will hopefully allow one million Americans to survive the impacts of both comets.

Of the one million persons, 200,000 have been preselected because of their skills/knowledge in areas important to restarting the nation after the disaster. The remaining 800,000 will be selected by lottery from among those under 50 years of age a couple of days before the comets strike. Those preselected or randomly selected will be taken by the military to the ark shortly before the comets hit. Martial law is imposed in order to maintain law and order.

Real Science. In the case of *Armageddon*, an object the size of Texas would be seen by many more than only 8 telescopes a mere 18 days before collision. If, for whatever reason, it was not widely observed until then, the fact that the word of its existence would then quickly get out would not jeopardize the mission to divide and deflect the asteroid because the space shuttles were already built and available. The effect on law and order of a planetary killer was not explored in the film. In reality, the human race would be doomed in this scenario. How the human race behaves in its last 18 days would probably vary widely among and within countries.

The scenario described in *Deep Impact* seems very realistic. First keep the knowledge of the impending collision from the public until a plan (in this case the building of the Messiah) was nearly complete so that a breakdown in law and order would not endanger the mission to stop the comet. Then be as optimistic as possible about the prospects of success for the Messiah mission. The likelihood of stopping the comets at the last minute by the firing of the Titan missiles is not high. They were not built to take down an asteroid or comet.

Chapter One: Sixty-Five Million Years and Counting

The building of an underground ark, use of a lottery to select most of its occupants, and imposing martial law to keep society operating while the ark was being completed, all seems very plausible, as does the recall of American troops from abroad since they would be needed to maintain order at home.

An interesting question is whether the sky would have cleared sufficiently after two years for the million people to emerge from the ark and begin to produce food by farming and the raising of livestock. The answer to that question is not clear. One other issue that was not raised in *Deep Impact* was having only one ark in one location. Since the exact location that the comet(s) would strike was not known until the last minute because the actions of either the Messiah or the Titan missiles could affect the impact point(s), it would have been prudent to have at least two large caves, just in case the comet happened to crash near one of them, killing all within by the direct blast of impact.

Chapter One: Sixty-Five Million Years and Counting

QUESTIONS

1. Search the Internet for the most recent information on 2004MN4, named Apophis. How close to the Earth is it expected to come on April 13, 2029?

2. To date what fraction of Near Earth Asteroids has had their future orbits plotted?

3. What is the value on the Torino scale (used to categorize the Earth's impact hazard associated with asteroids and comets) for a certain collision with an asteroid or comet capable of causing a global climactic catastrophe?

4. Another asteroid, 1950DA, which is 1,300 meters in diameter (much larger than the 320 meter diameter 2004MN4), is predicted to have a 1 in 300 chance of impacting the Earth, probably in the North Atlantic. What is the possible impact date for this object?

5. What evidence do geologists use to infer impacts in the geological record?

CHAPTER TWO

NUCLEAR TERRORISM

THE SCIENCE OF *THE PEACEMAKER* AND *THE SUM OF ALL FEARS*

NUCLEAR REACTORS

There are three general types of threats posed by terrorists using nuclear weapons or nuclear reactors. First, terrorists might attack a nuclear reactor. A nuclear reactor uses nuclear reactions in its core to produce heat which generates steam. The steam then turns a turbine which generates electricity in the same manner as a coal, oil, or gas powered commercial generator of electricity. The reactor core consists of rods of nuclear material, such as Uranium 235, which are placed in graphite moderators which slow down the fast neutrons emitted by the disintegrating U-235 nuclei. Slower neutrons are more likely to interact with other U-235 nuclei to split them and thus release energy.

Reactors also contain rods of cadmium, or boron steel, which can be moved into or out of the graphite to increase or decrease the absorption of neutrons thereby speeding up or slowing down the nuclear reaction and the amount of heat energy being produced in the reactor core. A reactor is a construction that produces a controlled "chain reaction." Reactors are heavily shielded to protect workers from radiation emitted by the radioactive elements in the core.

<u>Nuclear fission</u> is the process of breaking a large nucleus into two or more smaller nuclei. If the combined masses of the smaller

Chapter Two: Nuclear Terrorism

nuclei are less than that of the parent nucleus, the mass loss is converted into radiant energy, using the famous formula of Albert Einstein, $E = mc^2$, where E is the energy in Joules, m is the mass loss in kilograms and c is the speed of light in a vacuum, namely 300,000,000 meters per second. Thus c^2 is an enormous number, namely 9 followed by 16 zeros! This conversion of mass to energy in nuclear processes is the fundamental science behind both nuclear power and nuclear bombs.

In U-235 each disintegration also releases 2 or more neutrons. Neutrons are a fundamental component of nuclei which usually consist of both positively charged protons and electrically neutral neutrons. If these released neutrons strike other U-235 nuclei, the collision results in the disintegration of other nuclei with the release of yet more neutrons and energy. This results in a so-called "chain reaction" in which the two released neutrons each strike two other U-235 nuclei, which disintegrate releasing 4 neutrons, which strike four U-235 nuclei, which disintegrate releasing 8 neutrons, and so on. This all occurs in a small fraction of a second causing a nuclear explosion. However a chain reaction requires a sufficient mass of the U-235 so that neutrons released by the disintegration of one nucleus will strike other U-235 nuclei instead of leaving the sample without splitting another U-235 nucleus. This sufficient mass is called the "<u>critical mass.</u>" In an atomic bomb two or more sub-critical masses are brought together to form a critical mass and/or an amount of U-235 is explosively compressed in order to bring the U-235 atoms closer together to trigger a chain reaction.

In the fuel rods of a nuclear reactor there is not sufficient U-235 to lead to a chain reaction, so a nuclear explosion is impossible. However if the water which cools the rods is removed (by an accident or by sabotage such as detonating an explosive device next to the reactor core), then the nuclear rods quickly heat up and evaporate highly radioactive materials into the atmosphere. All American reactors are housed in a strong containment building which was built to withstand even the impact of the largest airplanes in existence when these reactors were commissioned. The purpose of the containment building is to prevent radioactive materials accidentally released from the reactor core from reaching the outside atmosphere.

FAILURE OF NUCLEAR REACTORS

THE THREE MILE ISLAND ACCIDENT

The worst commercial nuclear power accident in the United States occurred on March 29, 1979 at the Three Mile Island nuclear reactor near Harrisburg, Pennsylvania. The accident was caused by a malfunctioning pilot relief valve at the top of the water pressurizer which failed to close when the water pressure dropped. This created an opening in the primary water cooling system.

However, the indicator light in the plant's control room incorrectly showed that the valve was closed. The open valve caused a loss of cooling water for some two hours. A back up system then came on automatically to inject cooling water into the reactor, but the operators thought that too much water was being pumped into the system and reduced the high pressure injection from 1000 gallons a minute to 100 gallons per minute. This resulted in much of the reactor core being uncovered by water which then released radioactive material into the containment building. That building prevented nearly all of the radioactive materials from being released to the atmosphere.

However, a buildup of highly flammable hydrogen gas occurred in the containment building and authorities feared that a large hydrogen explosion might breach the containment walls. Thus tens of thousands of people were evacuated from the area, which indicated the seriousness of the accident. A Presidential Commission investigated the accident and concluded that "the training of TMI personnel was greatly deficient. While training may have been adequate for the

33

operation of the plant under normal circumstance, insufficient attention was paid to possible serious accidents."

CHERNOBYL

The world's worst nuclear accident occurred in the Chernobyl nuclear power plant in the former Soviet Union on April 26, 1986 when one or two explosions ripped apart a reactor and released huge amounts of radioactive materials into the atmosphere. The Chernobyl reactor did not have a containment building around the reactor core. Local firefighters bravely fought the fire and prevented it from spreading to other reactors in the complex. Many later died from the radiation exposure they received. The fire was eventually contained by using helicopters to drop 5,000 tons of sand and boron to plug the leak in the reactor. The reactor was eventually encased in 300,000 tons of concrete.

The release of radioactive materials into the atmosphere affected the health of over 350,000 people in parts of the Ukraine alone where people cannot safely drink water or eat locally produced food. The Chernobyl catastrophe demonstrates the amount of damage that can be done if the radioactive materials in a reactor core reach the atmosphere.

Since all of the 100 or so reactors in the United States have containment buildings it is highly unlikely that an accident would result in a Chernobyl-like disaster. However, reactors could be an inviting target for terrorists.

NUCLEAR TERRORISM

There are two dangers that must be guarded against. First, terrorists could try to destroy part of the containment building and also breach the reactor core itself, using either modern jumbo jets that the

Chapter Two: Nuclear Terrorism

containment buildings may not be able to prevent from penetrating or using a truckload of conventional explosives, similar to that used in the first attack on the World Trade Center.

The second target could be the spent fuel rods which are usually stored near to the reactor in pools of water. If the water were to be removed, these rods might start to evaporate their radioactive components into the atmosphere. There are reportedly about 40,000 tons of spent fuel rods in the United States.

Highly radioactive wastes from nuclear reactors may eventually be transported across the country to a permanent repository in Yucca Mountain, Nevada. It is presently scheduled to open in 2017. Terrorists might attempt to destroy the containers for these radioactive wastes in transport resulting in the release of significant amounts of radioactive materials to the atmosphere.

DIRTY BOMBS

The second type of terrorist nuclear threat involves the use of a so-called "dirty bomb." This bomb consists of a conventional explosive surrounded by radioactive materials. This is fairly simple to build and will produce damage both from the conventional explosive as well as the dispersal of radioactive materials. One report indicated that over 800 radioactive sources are missing from medical and research facilities in the United States alone. Thus it may be relatively easy for terrorists to obtain some radioactive materials, particularly from research facilities in the former Soviet Union. However, the detonation of this kind of device will not produce anywhere near the death and destruction of a nuclear bomb.

NUCLEAR BOMBS

This is the threat that most concerns both the public as well as government officials. In order to understand the threat we need to first discuss the components of a nuclear bomb. All such bombs consist of a core of either enriched U-235 or plutonium. Natural uranium consists of only about 0.7% U-235 and 99.3% of U-238. However U-238 does not undergo fission when struck by a neutron but rather absorbs the neutron. Thus naturally occurring uranium will not produce a chain reaction (and therefore a nuclear explosion).

The goal is to obtain Highly Enriched Uranium (HEU), i.e., uranium containing at least 20% of U-235. One needs to separate out the U-235 from the U-238 using such devices as gas centrifuges which moves the heavier U-238 to the outside of a drum rotated at high speeds while the lighter U-235 is extracted from the center of the rotating drum.

Alternatively one can convert the abundant U-238 into Plutonium 239 using so-called breeder reactors. The plutonium can then be used to power nuclear plants or can be used to construct nuclear bombs. Doing so will reduce the radioactive materials that must be stored for thousands of years since these plants use up more of the uranium than normal nuclear power plants. The plutonium can also be used to construct nuclear bombs.

A nuclear bomb detonates either by bringing together two sub critical masses of HEU or plutonium or by compressing HEU or plutonium using explosive panels that surrounds the nuclear core of the bomb. When the explosive panels detonate they bring the Uranium 235 or Plutonium 239 atoms close enough together to initiate a chain reaction.

Thus the construction of a nuclear bomb involves acquiring not only the knowledge to design the bomb and its explosive components but also the technological ability to produce (or acquire) either HEU or Plutonium 239. One report states that about 1,665 tons of HEU and 147 tons of plutonium are stored for military purposes at hundreds of

Chapter Two: Nuclear Terrorism

facilities worldwide. Another report states that about 40 countries have about 2070 tons of weapons usable fissile material (enough to produce more than 130,000 nuclear weapons).

How much HEU or plutonium is needed to build a bomb? About 50 kilograms of HEU for a simple designed nuclear bomb (only 12 kilogram of HEU for a more sophisticated bomb design) or about 4 kilograms of plutonium for a more sophisticated bomb design. A kilogram weighs about 2.2 pounds. Thus the amount of plutonium needed is the size of a baseball. One report states that between 1993 and 2004 there were 25 highly credible reports of theft of HEU or of plutonium. A total of 39 Kg of HEU or plutonium were intercepted during transit, sale, or diversion attempts over this period of time.

Of course there may well be other thefts that are unknown to government officials. The greatest concern has been over the nuclear scientists in the former Soviet Union. At one time they were the elite of Soviet Society. Now, with decreased funding, low wages delayed sometimes for months, and bleak prospects, the temptation to sell either their expertise or the fissionable material to which they have access is of concern to anti terrorist experts. In all credible thefts of weapons usable materials known to date, the material was stolen by insiders.

Note that HEU bombs are easier to make and are much less radioactive than plutonium bombs, making HEU bombs safer to handle.

Unfortunately, there are diagrams of nuclear bombs readily available on the Internet. For example, one can find crude diagrams of both a gun assembly atomic bomb (like the Hiroshima bomb) as well as an implosion bomb in the online free encyclopedia, Wikipedia, (its URL is http://www.wikipedia.com) as well as on many other web sites. Thus the main deterrence to terrorists making a nuclear bomb is the difficulty in obtaining the needed amounts of either HEU or

37

Chapter Two: Nuclear Terrorism

plutonium.

Finally, terrorists might steal, or buy, an entire nuclear weapon from individuals working in the nuclear programs of the nine nations believed to presently have nuclear weapons, namely, the United States, Russia, Great Britain, France, China, India, Pakistan, Israel and North Korea.

This is what happens in the plots of both films we will discuss next. The National Resources Defense Council estimated that as of 2002 there were 22,000 nuclear weapons in the world. The largest thermonuclear bomb ever produced (called the "Tsar Bomba" by the Soviets) had a yield of 100,000,000 tons of TNT and the smallest nuclear weapon (called the "Davy Crockett") was built by the United States and had a yield of only 100 tons of TNT; it weighed only 51 pounds.

Thermonuclear bombs have an atomic bomb at their core to raise the temperature of hydrogen to millions of degrees so that in a multi step process four hydrogen nuclei (i.e. protons) are combined into one helium nucleus with the release of energy. This is the process which powers our sun and other stars. It does not occur at room temperature since positively charged protons electrically repel one anther. These protons must be moving at high speeds for the attractive short range nuclear force to come into play to then bind them into a helium nucleus. For this to happen the hydrogen must be at very high temperatures. This is why an atomic bomb is used to create these high temperatures. There is no limit to the size of a thermonuclear ("hydrogen") bomb. However it takes the resources of a major state to produce one so there is little chance that terrorists could build one.

Chapter Two: Nuclear Terrorism

FILM SUMMARIES

THE PEACEMAKER

Dream Works (US), 1997, color, 124 minutes

Cast: George Clooney (Colonel Thomas Devoe), Nicole Kidman (Dr. Julia Kelly), Armin Mueller-Stahl (Dimitri Vertikoff) and Alexander Baluyev (General Aleksandr Kodoroff)

Director: Mimi Leder

The movie opens at a baptism in Yugoslavia. A minister attending the service is called outside to answer a phone and is killed by an assassin. We later learn that this was done so as to replace him with one of the terrorists. The film then depicts the loading of a missile with 10 warheads onto a train at a Russian missile base in Chelyabinsk. The missile is headed to a destination where the warheads will be deactivated as part of a reduction in Russia's nuclear arsenal. The train is under the command of a nervous Russian officer. A forbidding looking Russian general sees the train off.

Several hours later the soldiers on the warhead train are asleep when a second train pulls alongside and troops from the second train kill all of the soldiers on the warhead train except for the nervous officer, who is in league with the killers. Nine of the warheads are loaded onto the second train and the tenth is set to detonate in a matter of minutes. The second train pulls into a tunnel to be protected from the blast which occurs only 15 kilometers away. The explosion occurs just a few minutes after the train carrying the warhead collides head on with a passenger train. The warhead yield is about 700 kilotons of TNT.

The film moves next to Washington, DC, where we meet Dr.

39

Chapter Two: Nuclear Terrorism

Julia Kelly, formerly of the Jet Propulsion Laboratory, who is the acting chief of the nuclear smuggling section. She asks for someone familiar with the Russian military and is assigned Colonel Thomas Devoe, who is seen testifying before a Congressional committee about the methods he has employed in Russia (including bribery) to keep weapons of mass destruction out of the hands of terrorists.

Devoe points out that the satellite imagery of the trains before the collision suggests that the soldiers on the train carrying the warheads were already dead before the collision occurred. He further believes that the nuclear explosion was a ruse to conceal the theft of the other nine warheads for sale by the thieves. A bit of detective work leads Devoe and Kelly to suspect that Russian General Kodoroff is the brains behind the robbery. Although he supposedly perished in the explosion, they trace phone calls to his mistress to confirm that he lives.

Devoe then arranges to meet his Russian contact, Dimitri Vertikoff, in Europe where they obtain trucking schedules for days before the theft from a business magnate who controls most trucking operations in Russia. Vertikoff is killed in a firefight after the schedules are obtained.

From the schedules they identify one truck heading towards the Iranian border, with, they believe, 9 warheads and General Kodoroff. The truck is identified by a ruse in which Devoe convinces Kodoroff over his cell phone that there is a smart bomb falling towards him, causing him to pull his truck out of a long line of trucks so that its license plate can be seen by the reconnaissance satellite.

After a bloody chase in which an American helicopter is shot down by Russian ground to air missiles for violating Russian air space, US Army Rangers, led by DeVoe, capture the truck, and kill Kodoroff, only to find that one warhead is missing. It has been sold to a Yugoslav terrorist who blames the West for the wars in his country, and the death

Chapter Two: Nuclear Terrorism

of his wife and daughter. The warhead has been modified by a Pakistani nuclear scientist into a portable nuclear weapon that can be carried as a backpack. The captured Pakistani scientist says that the weapon will have an explosive yield of one or two kilotons of TNT.

A tracing of payments to the Russian General leads to the Yugoslav who has the weapon, but he has already departed for the US when allied troops storm his home. He is part of a peace delegation to the United Nations and the bomb is hidden in a diplomatic pouch he is carrying. Thus the closing of all airports in New York City by a presidential nuclear alert declaration does not keep the nuclear bomb out of the city.

Since authorities know that the United Nations is the target, helicopters with sensitive radiation detectors sweep over Manhattan while massive numbers of police and military try to intercept the bomb. Even though the terrorist team is located and both members are killed in shootouts, the bomb is counting down to detonation in less than 3 minutes when they open it. Dr. Kelly only has time to knock off one of the explosive plates surrounding the plutonium core before the explosives detonate. It is sufficient to prevent a nuclear explosion and both Kelly and Devoe survive the explosion.

MOVIE TRIVIA

After Devoe introduces Julia Kelly, the Russian says to Devoe (in Russian) "A beautiful woman and a Ph.D. You're way out of your league." Then Kelly replies in Russian "You have no idea, Colonel."

When the hijackers short wires together to divert the bomb train into a head-on collision, the track switch appears to move in the wrong direction.

According to Wikipedia, the character of Dr. Julia Kelly is

41

loosely based on Harvard professor and terrorism scholar Jessica Stern, who served as director for Russian and Eurasian Affairs at the United States National Security Council.

THE SUM OF ALL FEARS

Paramount Pictures (US), 2002, color, 126 minutes

From the novel by Tom Clancy

Cast: Ben Affleck (Jack Ryan), Morgan Freeman (William Cabot), James Cromwell (President Robert Fowler), Ciaran Hinds (President Nemerov), Bridget Moynahan (Dr. Catherine Mueller) and Liev Schrieber (John Clark)

Director: Alden Robinson

The movie opens in 1973, during the Yom Kippur War, when an Israeli A-4 jet carrying a nuclear bomb is shot down over the desert somewhere in the Middle East. In 2002 the bomb is found and sold to an arms dealer named Olson who in turn sells it to a neo-Nazi named Richard Dressler for 45 million dollars.

In Russia a new President, Alexander Nemerov, takes office after the sudden death of President Zorkin. Nemerov is viewed in America as a hard liner. Director of Central Intelligence William Cabot seeks the views of young CIA analyst Jack Ryan, who has done research on Nemerov. Cabot and Ryan go to Russia to inspect one of Russia's nuclear weapons facilities. They are unexpectedly invited to the Kremlin to meet with Nemerov personally. The Russian President wants Cabot to deliver a message to the US President: Do not intervene in Chechnya.

During the inspection of the nuclear facility, Ryan notices that

Chapter Two: Nuclear Terrorism

three scientists are missing. He is told that one scientist is out sick, a second is on vacation and a third deceased. However, Cabot's covert informant in Moscow, code named "Spinnaker," reports that the Russians do not know the whereabouts of the three missing scientists, all of whom are nuclear bomb experts.

Upon returning to the United States, Cabot sends CIA operative John Clark to find the missing scientists. Clark tracks the missing scientists to a facility in the Ukraine. Meanwhile an unauthorized gas warfare attack is launched against Grozny, the capital city of Chechnya. Although President Nemerov takes responsibility for the attack, Ryan believes it was not authorized by the Russian President. President Fowler responds by sending NATO peacekeeping troops to Chechnya.

The film also depicts the arrival of the nuclear bomb in Baltimore, Maryland, disguised as a cigarette vending machine. Dressler hopes to get America and Russia to fight each other. Detonating a nuclear bomb in Baltimore is part of his plan.

Meanwhile, Ryan and Clark return to the Ukrainian facility to discover that the three scientists are all dead. They surmise that the scientists had constructed a working bomb for Dressler. They know that the bomb is large and heavy and are able to trace a shipment of that size from the Ukraine to Baltimore.

Ryan then tries to reach Cabot to tell him that the bomb is in Baltimore, but President Fowler and Cabot are attending a football game in the very stadium where the bomb is planted. The noise in the stadium makes it difficult for Cabot to hear Ryan. When he does, Cabot orders the Secret Service to rush the President out of the stadium. The bomb explodes moments afterwards with an explosive force much less than that of Hiroshima, but still causing huge casualties.

Chapter Two: Nuclear Terrorism

President Fowler is rescued by Marines from his badly damaged car and then is taken to a Boeing Advanced Airborne Command Post with his cabinet. They suspect that the bomb is Russian. Ryan and his girlfriend, a physician, survive the blast, but Cabot dies later in a makeshift hospital.

After learning of the explosion, Dressler orders a Russian Air Force general, who was in his pay, to send Russian Tu-22M backfire bombers to strike a US aircraft carrier after telling the pilots that a US ICBM has hit Moscow. The strike badly damages the ship and President Fowler orders Air Force F-16s to attack the Russian air base from which the bombers came. Tensions rapidly mount as Fowler orders SNAPCOUNT, which brings the military to maximum readiness, preparing to launch an all out nuclear attack against Russia. Seeing that the US has dispatched B-2 stealth bombers and nuclear submarines, Nemerov prepares to launch his ICBMs against the US.

Meanwhile, Ryan has discovered from the Air Force Radiation Assessment team that the plutonium for the Baltimore bomb was manufactured in our Department of Energy "K Reactor" in Savannah River, South Carolina in 1968, thus indicating the nuclear bomb was originally American, not Russian. He also discovers that Dressler was behind the attack and using Cabot's text messenger he learns from Spinnaker that the US had secretly sent the plutonium to Israel for its nuclear weapons program.

Ryan gets to the Baltimore docks to find Dressler's American contact (who unloaded the bomb and placed it in the stadium) murdered by Dressler's South African assassin. Ryan captures the assassin but is unable to force him to talk before Maryland State Police arrive. He next gets to the Pentagon, where he is able to communicate the truth to both Presidents Nemerov and Fowler. Relying on Ryan's word, both leaders stand down.

The two Presidents later meet and make peace as agents of

Chapter Two: Nuclear Terrorism

both governments hunt down and kill Dressler, the Russian Air Force general who launched the strike against the US carrier, and Olson. The final scene takes part in Washington, D.C., where the two Presidents address the Baltimore tragedy and the future of nuclear weapons. In a nearby park, Ryan and his girl friend, Dr. Cathy Mueller, are having a picnic when they are approached by Grushkov, who is revealed to be Spinnaker. The Russian gives Dr. Mueller a "modest gift" for her engagement to Ryan. When Ryan asks how Grushkov could possibly have known of their engagement since they had not told anyone, he simply smiles and walks away.

MOVIE TRIVIA

In the book on which the film is based, Jack Ryan is Deputy Director of the CIA, while in the film he is a low level intelligence analyst.

In the book, Ryan is already married to Catherine Mueller and they have children while in the film he is romancing her.

In the book, CIA Director Cabot is a political appointee who is not very good at his job while in the film he is very skilled and is sorely missed in dealing with the crisis after his death.

SCI-FI SCIENCE vs. REAL SCIENCE

SOURCE OF THE NUCLEAR MATERIALS IN THE BOMBS

THE PEACEMAKER. The plutonium comes from a Russian nuclear missile which was to be deactivated.

THE SUM OF ALL FEARS. The plutonium comes from an

Israeli bomb that was sent from the US.

Real Science. *The Peacemaker* scenario is of real concern to experts, namely that a corrupt Russian military officer might be able to steal a plutonium bomb core.

As far as is known, the US did not send bomb grade plutonium to Israel. Furthermore, the idea that if Israel intended to use a nuclear bomb in the Yom Kippur War it would deliver it using only one aircraft without other planes to protect the bomb carrying plane is completely unrealistic. In addition, if the plane was shot down, the Israelis would surely then have sent in ground forces to recover the bomb because it could be used against Israel. Thus *The Sum of All Fears* scenario of the origin of the bomb is not believable.

FASHIONING OF PLUTONIUM INTO A USABLE BOMB

THE PEACEMAKER. The backpack bomb is constructed by a Pakistani scientist.

THE SUM OF ALL FEARS. The much larger bomb is built by three Russian nuclear scientists.

Real Science. Both plots are plausible. Russian nuclear scientists are receiving low pay and might be receptive to offers of large sums of money to use their technical expertise. Pakistan's nuclear technology has been in part exported to other countries. However, it is not plausible that the Pakistani physicist could completely reconfigure the bomb in the back of a truck over a period of only hours using only elementary tools. It is more plausible that three Russian scientists could rebuild a bomb in a building with extensive equipment over a period of many days or weeks.

Chapter Two: Nuclear Terrorism

SIZE OF THE BOMB AND THE EXPLOSIVE YIELD

THE PEACEMAKER.. The bomb is in a backpack that can be carried by one person and has a yield of a couple of kilotons.

THE SUM OF ALL FEARS. The bomb is housed in a large cigarette vending machine and has a yield of perhaps ten kilotons.

Real Science. Both pictures are realistic. A bomb with a larger yield would be too large to carry.

Whether a bomb as small as that seen in *The Peacemaker* would have a yield of as much as 2,000 tons of TNT is not known to the authors.

DETAILS OF THE INSIDE OF THE BOMB

THE PEACEMAKER. The use of explosive panels to compress the plutonium is accurate. Hence the knocking off of one explosive panel to avoid compressing the bomb when the timer fires seems very logical. This channels at least part of the explosive force outwards and not inwards to the plutonium core.

THE SUM OF ALL FEARS. We are not shown the inside structure of the bomb.

RADIATION EMITTED FROM THE BOMBS BEFORE DETONATION

THE PEACEMAKER. Since plutonium is highly radioactive, the bomb would be giving off radiation and this radiation would be detectable from the air as depicted in the film. This radiation might be lethal to whoever was carrying the bomb, but since this was a suicide mission that would not deter the terrorists. It is not clear whether the terrorist who carried the bomb from the truck to the meeting with his brother (the ambassador) would have become ill from the effects of the radiation before reaching New York City.

THE SUM OF ALL FEARS. Since the bomb is inside a vending machine it could be shielded to reduce the amount of radiation being emitted by the device, thereby making it harder to detect the plutonium or uranium core inside of it.

THE EFFECT OF NUCLEAR BOMBS DETONATING

THE SUM OF ALL FEARS. . Much of Baltimore is destroyed and the President's motorcade badly damaged even though it is some distance from the football stadium when the blast occurs. The film states that the yield was much less than that of the Japanese blasts (Hiroshima about 11 kilotons and Nagasaki about 21 kilotons).

Consider the probable damage from a bomb with half the destructive power of the bomb dropped on Nagasaki. According to one source, a 10 kiloton bomb would completely destroy all structures within a 1/3 mile radius with 100% fatalities. There would also be fatal radiation to all exposed to the blast within a 3/4 mile radius, with severe damage to all buildings. Everything within a one mile radius would sustain major damages, and inhabitants would be exposed to significant radiation and fires. Many others would die later from the fallout, depending on the wind direction after the blast. One estimate of the total deaths from a bomb of that size exploded in the center of Washington D.C. was as high as 700,000. Thus the film's depiction of widespread destruction in Baltimore seems reasonable.

THE PEACEMAKER. The detonation is non nuclear. The church partially contains the radioactive materials. However, the windows are blown out and one would expect that Dr. Kelly and Colonel Devoe would have been injured by shrapnel containing highly radioactive materials, leading to permanent injury or even death from the radiation received by them.

IDENTIFYING THE SOURCE OF THE NUCLEAR MATERIAL

THE PEACEMAKER. The authorities know that the missiles on the train are Russian made so this is not an issue.

THE SUM OF ALL FEARS. The assertion that the Air Force Radiation Assessment Team would know that the radiation profile lingering after the explosion would allow them to identify where the plutonium came from is not plausible. They may be able to identify from the lingering radiation that the bomb contained plutonium rather than enriched U-235, but that is about all that could be identified, not the site and year the plutonium was produced from U-238. The speed with which they examine the radiation and come up with its source (a matter of minutes) is also pure Hollywood.

THE TERRORISTS

THE PEACEMAKER. The terrorists are from the former Yugoslavia, probably ethnic Muslims, although that is not certain.

THE SUM OF ALL FEARS. The terrorists are Neo-Fascists. In the book on which the film was based, they are Muslim extremists. The film was completed months before 9/11/2001 so this change was not in any way a reaction to the September 11 attacks, but rather was done purely for the elements related to the plot, as the Director stated that Muslim extremists would not be able to plausibly accomplish all the elements in the storyline. In addition, the Council on American -Islamic Relations (CAIR) had undertaken a two year long lobbying campaign (which ended in early 2001) against using "Muslim villains"

Chapter Two: Nuclear Terrorism

in films as the original book version had done. Director Alden Robinson is reported to have assured CAIR that he had no intention of promoting negative images of Muslims.

SATELLITE IMAGERY

THE PEACEMAKER. The images received from "spy satellites" are utilized repeatedly in the film. In the beginning of the film we are shown images of the trains just before they collide. Later we see images of the truck carrying the nine nuclear bombs. The film emphasizes the need to intercept the truck while it is still being tracked via a spy satellite. In a matter of one hour and 46 minutes, it will be lost from view and then will be impossible to intercept. Thus helicopters are dispatched to intercept the truck even though it means violating Russian air space.

Satellites circle the earth over different periods of time. For example a geosynchronous satellite is one that circles the Earth every 24 hours, exactly the same time as the Earth takes to rotate once on its axis. Hence a geosynchronous satellite remains above the same points on the Earth at all times. These satellites are used for transmitting television signals, cell phone communications, etc.

However, to remain in this orbit it is about 22,000 miles above the surface of the Earth. This is too far away to be useful as a "spy satellite" (officially called "reconnaissance satellites"), which needs to be much closer to the surface of the Earth in order to get more detailed images of installations, vehicles and even people. The closer the satellite is to the Earth's surface, the shorter is the time to make a complete orbit. These satellites are believed to typically circle the Earth at a height of 60-120 miles or so, taking perhaps 90 minutes. That means that a "spy satellite" will only be able to "observe" a given point on the Earth for a short part of its orbit, lasting typically only a few minutes.

Chapter Two: Nuclear Terrorism

The resolution of modern spy satellites is classified: They are believed to be able to see objects as small as 6 inches. This is sufficient to identify any vehicle, such as the truck in *The Peacemaker*, but is not sufficient to read the headlines of a newspaper as some in the popular media have suggested.

Therefore in *The Peacemaker* script when the audience is told that the truck carrying the missiles will be in observation for 1 hour and 46 minutes, done by a satellite that can read the license numbers on the truck: this is a gross exaggeration. To get that kind of resolution the satellite would have to be so close to the Earth that it would circumnavigate the globe in less than 1 hour and 46 minutes and thus could not possibly remain above one location on Earth for that period of time.

Chapter Two: Nuclear Terrorism

QUESTIONS

1. How many deaths were caused by the Hiroshima and Nagasaki bombs?

2. In *The Peacemaker*, why didn't Dr. Kelly just start cutting all the wires she could see in the bomb in order to stop it from exploding?

3. In *The Sum of All Fears*, would Jack Ryan likely have been exposed to significant amounts of radiation when the bomb detonated in Baltimore?

4. Do you think that the resolution of the confrontation between the United States and Russia in *The Sum of All Fears* is plausible? Why or why not?

5. In *The Peacemaker*, is the shootout between DeVoe and three carloads of gangsters plausible? Why or why not?

6. What kinds of security measures exist on ports of entry to the United States to interdict a nuclear weapon as it enters the country?

7. Do you favor increasing the number of nuclear power plants in the United States? Why or why not?

8. What are the measures the United States and other countries can take to minimize the chances that terrorists will obtain nuclear bombs?

9. List three web sites that address nuclear terrorism.

10. What kind of nuclear material was used in the bombs dropped on Hiroshima and Nagasaki? Was it enriched uranium or plutonium?

CHAPTER THREE

GLOBAL WARMING AND THE GREENHOUSE EFFECT

THE SCIENCE OF *THE DAY AFTER TOMORROW* AND *AN INCONVENIENT TRUTH*

The greenhouse effect is another name for global warming caused by water vapor, carbon dioxide and other gases in Earth's atmosphere. Without the thermal warming due to the natural greenhouse effect, the Earth's climate would be at least 59 Fahrenheit degrees cooler than it is, too cold for many organisms to survive. On Mars, with its thin atmosphere and therefore virtually no greenhouse effect, the mean surface temperature is -80 Fahrenheit degrees.

WHAT IS THE GREENHOUSE EFFECT?

The greenhouse effect results from the interaction between sunlight and the layer of greenhouse gases in the Earth's atmosphere that extends up to about 60 miles above the surface of the Earth. Sunlight consists of a range of short wavelength electromagnetic waves most of which easily pass through the atmosphere and reaches the Earth (unless reflected by atmospheric particles, such as sulphur from volcanic explosions).

About 25%-30% of the Sun's radiation is reflected back into space by clouds and atmospheric particles. Another 20% is absorbed by the atmosphere. For example ozone absorbs most of the Sun's dangerous ultraviolet radiation and prevents that radiation from reaching the surface of the Earth. Somewhat over 50% of the Sun's

Chapter Three: Global Warming and the Greenhouse Effect

energy, mainly in the visible light spectrum, passes through the atmosphere and reaches the surface of the Earth. Of that 50%, approximately 85% is absorbed by the Earth while the remaining 15% is reflected back into the atmosphere from reflective surfaces such as snow, ice, and sandy deserts.

Some of the 85 % of the Sun's radiation that is absorbed by the Earth is re-radiated back at wavelengths in the infrared portion of the electromagnetic spectrum. Certain gases in the atmosphere such as water vapor, carbon dioxide, methane and nitrous oxide absorb this infrared radiation and thus prevent it from escaping into outer space. These gases in turn re-radiate the heat in all directions.

Some of this heat returns to the Earth to further warm its surface in what is called the greenhouse effect, while the remainder of the heat is eventually released into outer space. These processes creates an equilibrium between the total amount of heat that reaches the Earth both directly from the Sun as well as from the greenhouse gases and the amount of heat the Earth re-radiates into outer space. The more net heat that remains on the Earth, the warmer the Earth will become.

The greenhouse effect gets its name from the similar manner in which a glass greenhouse works. Sunlight passes through the glass of the greenhouse to warm its interior whereas the infrared energy that is re-radiated from the inside of the greenhouse is absorbed by the glass and some of this absorbed energy is re-radiated back into the greenhouse by the glass. In the planetary model, the greenhouse gases perform essentially the same function as the glass cover performs for a greenhouse. However, much of the effectiveness of greenhouses comes from the fact that the air inside a greenhouse is confined and therefore cannot rise into the atmosphere when it is heated. Thus greenhouses work primarily by preventing the motion of air (called convection) whereas the atmospheric greenhouse effect works primarily by reducing the radiation loss from the Earth into outer space. Note that automobiles in the summer can be very effective (and even deadly) greenhouses when cars are left closed with all windows tightly shut. This has resulted in tragedies for pets and small children left inside such cars.

TYPES OF GREENHOUSE GASES

The atmosphere consists mainly of nitrogen (78%) and oxygen (21%). The greenhouse gases make up less than 1% of the atmosphere.

The most abundant naturally occurring greenhouse gas is water vapor, which accounts for most of the natural greenhouse effect. Human activity does not directly affect the water vapor levels in the atmosphere. However, as human activities increase the concentrations of other greenhouse gases in the atmosphere, this may produce warmer temperatures, leading to greater evaporation from oceans, lakes and rivers and even from plants and soil, thereby increasing the amount of water vapor in the atmosphere and enhancing the greenhouse effect.

The next most abundant greenhouse gas is carbon dioxide, which circulates in the environment through a number of natural mechanisms known as the carbon cycle. The decay of plant and animal matter as well as volcanic eruptions release carbon dioxide into the atmosphere. Animals exhale carbon dioxide formed during the break down of food to nourish cells in the animal. Plants, on the other hand, absorb carbon dioxide from the atmosphere. Through photosynthesis, plants use this carbon dioxide to make their own food and release oxygen back into the atmosphere. One model of the atmosphere estimates that water vapor constitutes 36% of the greenhouse effect and that carbon dioxide contributes 9%, but the removal of both of these constituents from the atmosphere would lead to a greater reduction in the greenhouse effect than the sum of amounts that each separately contributes to the greenhouse effect.

Chapter Three: Global Warming and the Greenhouse Effect

Human activity is now producing carbon dioxide at a pace faster than Earth's natural processes can absorb it. These processes involve the use of fossil fuels such as oil, coal and natural gas, wood or wood products, and some solid wastes. When these products are burned to fuel electric power plants, power automobiles and heat buildings, carbon dioxide is released. In addition, humanity is cutting down large sections of forests and jungles, which both releases the carbon stored in the trees as well as diminishes the capacity of the remaining forests to absorb carbon dioxide.

Scientists have studied the gases trapped in ice cores going back thousands of years. Ice cores from Greenland and the Antarctic date back as long as 650,000 years ago. The concentration of carbon dioxide in these ice cores have never been greater than 300 parts per million (ppm) until very recent times. Carbon dioxide increased from about 274 ppm in pre industrial times (about 1750) to about 356 ppm in 1992, then to 367 ppm in 1999 and then to 379 ppm in 2005. Scientists are also able to estimate global temperatures from the gases trapped in the ice. Global surface temperature seems to broadly follow carbon dioxide in the atmosphere. Global temperature has increased by about 1 degree Fahrenheit since the 1860's.

Methane is another greenhouse gas which is produced in the decomposition of carbon containing substances in oxygen free environments such as wastes in landfills, in the digestive tracts of cattle, by microorganisms that live in damp soils such as rice fields, and by coal mining and other fossil fuel production operations. The abundance of methane has increased from about 0.7 parts per million in pre-industrial times to 1.7 parts per million in 1992. Although the concentrations of methane are far lower than that of carbon dioxide, molecule per molecule, methane is 20 times more efficient at trapping infrared radiation radiated from the surface of the Earth than is carbon dioxide.

Nitrous oxide is yet another greenhouse gas. It is released by

Chapter Three: Global Warming and the Greenhouse Effect

the burning of fossil fuels. Automobile exhaust is therefore a major source of this gas. Also, when nitrogen containing fertilizer breaks down in the soil, they emit nitrous oxide into the air. Nitrous oxide has increased by about 17% since 1750. Molecule for molecule, it is 300 times more efficient than carbon dioxide in trapping heat. Also, it can stay in the atmosphere for a century whereas methane only stays in the atmosphere for a decade or so.

Some of the most potent greenhouse gases are not produced by natural processes but are solely products of human activities. Fluorinated compounds such as CFCs, HCFCs, and HFCs are used in various manufacturing and cooling processes and are thousands of times more effective in trapping heat than carbon dioxide. Also, CFCs rise up to the upper atmosphere where ultraviolet radiation breaks down the CFCs, releasing chlorine which attacks the protective ozone in the upper atmosphere. HFCs are the safest substitute for CFCs since they do not contain chlorine and remain in the atmosphere for only a short time. However all of these compounds are greenhouse gases. Since CFCs destroy the protective ozone layer, their manufacture and use has been essentially banned world wide.

A recently identified addition to greenhouse gases is called <u>trifluoromethyl sulphur pentafluoride</u>. Although it exists in extremely low concentrations (almost nothing in the 1960s to about 0.12 parts per trillion in 1999), it is the most powerful trapper of heat ever seen. This is also not a naturally occurring substance.

PROJECTIONS OF THE EFFECTS OF GLOBAL WARMING

There is no doubt that the greenhouse gases contribute to the warming of the planet. What is hotly debated is the magnitude of the warming. Other processes contribute to the average temperature of the planet; these other processes have produced both warming and cooling

Chapter Three: Global Warming and the Greenhouse Effect

of the planet in the past. Over the last 100 years the average surface temperature of the planet has increased by over one degree Fahrenheit. However, most of the increase took place before 1940 whereas most of the human emissions of greenhouse gases occurred after 1940.

It may be that other natural processes masked the growth in the greenhouse effect. For example, sulphur haze cools the planet by reflecting sunlight back into the atmosphere before it reaches the ground. Sulphur emissions come from many of the same factories that emit carbon dioxide. In addition, volcanic eruptions can hurl huge amounts of sulphur into the upper atmosphere. Sulphur that reaches the upper atmosphere tends to remain there much longer than when it is confined to the lower atmosphere.

The effect of the eruption of Mount Pinatubo, in the Philippines, in 1991, was to hurl sufficient sulphur into the atmosphere to cause a global cooling of about one degree Fahrenheit for several years. This would obviously mask greenhouse warming for a while. In the long run, greenhouse warming overcomes sulphur cooling because most greenhouse gases will remain in the atmosphere for hundreds of years whereas human-produced sulphur emissions (which remain at lower altitudes) will fall to Earth in days or weeks. Also, the greenhouse gases warm the planet 24 hours a day whereas the sulphur acts to cool the Earth by reflecting sunlight only during daylight hours.

Climatologists have developed various computer-generated models to predict the effects of further increases in greenhouse gases. However, there is still debate about whether these models accurately reflect the complicated interactions between the land, the oceans and the atmosphere. Among the issues raised by skeptics is the fact that the greenhouse effect could start a cycle in which more clouds are formed, stopping the Sun's energy from reaching the Earth's surface in the first place and thus causing global cooling, not global warming. A 2001 Imperial College of London study comparing satellite data from 1970 and 1997 looked specifically at infrared readings from the Earth's

Chapter Three: Global Warming and the Greenhouse Effect

surface that reached outer space.

The lead investigator of the study concluded that the decrease in infrared radiation reaching outer space in 1997 compared to 1970 left no doubt that "the greenhouse effect is operating and what we are seeing can only be due to the increase in the gases." But he also said, "The effect of clouds on the planet is very complex, and frankly we don't understand it."

In 1988, the United Nations Environment Program and the World Meteorological Organization established the Intergovernmental Panel on Climate Change (IPCC) to provide periodic comprehensive assessments on climate change to guide policy makers around the world. In 1995, the IPPC issued a Second Assessment Report which included the following findings:

1. Human activities are increasing the atmospheric content of carbon dioxide and other greenhouse gases.

2. The surface temperature of the Earth has increased by 0.5 to 1.0 degrees Fahrenheit over the last century.

3. The predominance of evidence suggests that there is a measurable human influence on global climate.

4. Unless there is a reduction in the growth of greenhouse emissions, the Earth's average temperature is expected to rise by 2 to 6.5 degrees Fahrenheit by the year 2100. The sea level will rise by 6 to 38 inches by the year 2100. Further IPPC reports have supported these conclusions.

There is also other scientific information that is consistent with the occurrence of global warming. For example, the Greenland icecap is melting. The total amount of ice in the Greenland icecap is about 2.5 million cubic kilometers of ice, or about 10% of the ice in Antarctica.

Chapter Three: Global Warming and the Greenhouse Effect

However, the ice in Greenland is on land (unlike some ice that is already on the ocean), and any melting of it affects sea levels worldwide. Were the entire Greenland ice cap to melt, the sea surface water levels worldwide would rise by 22 feet or more. Even a partial melting of the polar and Greenland icecaps, plus the expansion that water undergoes with increasing temperature, could raise the global sea levels by perhaps two feet. This would flood large regions of agricultural lowlands in China, India and Bangladesh, causing widespread famine. Large parts of Florida and Louisiana could also end up under water. If the Antarctic ice were all to melt water levels would rise by 200 feet or more.

Rising ocean temperatures have been predicted to result in stronger (not necessarily more) hurricanes, because hurricanes gain intensity as they pass over warmer waters. Katrina and other hurricanes in the United States and elsewhere have been unusually powerful during the 2005 and 2006 hurricane seasons.

It is important to note that global temperatures do not rise uniformly over all parts of the Earth. For example, a 5 degree Fahrenheit increase in global temperature would probably result in the temperature at the poles increasing by 12 degrees while at the equator the increase might be only 2 degrees. This is in part because ice reflects about 90% of sunlight. However, when the ice at the poles melt, the water produced reflects only about 10% of the sunlight. The rest is absorbed by the water increasing local temperature considerably.

A worst case scenario is one in which a number of positive planetary feedback mechanisms amplify the greenhouse warming. For example, as the world warms, the oceans are less able to absorb carbon dioxide since carbon dioxide is less soluble in warmer water. A decrease in the creation of phytoplankton in the oceans could follow, caused by a reduction in the deep water nutrient supply to the upper levels of the ocean which are more stable than usual due to the effects of the warming.

As temperatures continue to rise, the tundra becomes a vast wetlands, releasing enormous amounts of the potent greenhouse gas methane. We have already seen the permafrost over parts of Alaska unfreezing. Thirty years ago the permafrost allowed trucks to drive to Alaska along frozen roads for an average of 275 days per year. In 2005 this was reduced to only 75 days per year. Next more high altitude clouds form containing ice crystals which help to trap even more heat. This leads to a runaway temperature increase. In this worst case scenario, the warming could not be reversed by any actions we could take and the human race would be doomed.

A runaway greenhouse effect using carbon dioxide and water vapor may have taken place on the planet Venus. Today there is little water vapor on Venus, perhaps because when Venus heated up, the water vapor may have risen very high into the atmosphere where the molecules of water vapor were split into hydrogen and oxygen by the ultraviolet light from the Sun.

Venus is much closer to the Sun than is the Earth. Hydrogen then escaped from the atmosphere while the oxygen recombined with other atoms. Carbon dioxide is the dominant gas in the present atmosphere of Venus which is hot enough to melt lead!

GLOBAL DIMMING

Global dimming is the gradual reduction in the amount of global direct sunlight reaching the earth's surface that was observed after the start of systematic measurements in the 1950's. This reduction is estimated to have been about 4% from 1960-1990 and has served to produce global cooling which has countered the global warming described above.

It is thought to have been caused by an increase in particulates (such as sulphate aerosols) in the atmosphere caused by human activities. The global dimming trend ended in the 1990's as global aerosol levels started to decline. This decline resulted in more sunlight reaching the surface of the Earth rather than being reflected by the particulates in the air.

As we improve air quality, that improvement could lead to a greater rate of increase in global warming than would be predicted from the greenhouse effect alone.

KYOTO PROTOCOL

In December, 1997 representatives from 160 countries attended an international conference in Kyoto, Japan to address the issue of global warming. The Kyoto Protocol, as the agreement is called, establishes goals for some of the countries to reduce their emissions of carbon dioxide. Under the Kyoto Protocol, the United States is mandated to reduce its carbon dioxide emissions to 7% below 1990 levels by the year 2012. Developing countries were exempted from greenhouse gas emissions restrictions. The United States Senate, which must ratify any treaty, strongly opposed some of the terms in the Protocol and it was never ratified.

One of the concerns raised about the Kyoto Protocol is its possible impact on the American economy, since some economic projections estimate that US emissions of carbon dioxide in 2012 will increase by about 34% above 1990 levels based on a "normal" growth rate for them. If we had to decrease them by 7% + 34% below this projected increase, some experts predict that the result would be an economic disaster with much higher energy prices and the loss of millions of jobs. Thus the discussion of global warming and the greenhouse effect is closely linked to economics and politics.

Chapter Three: Global Warming and the Greenhouse Effect

POPULATION EFFECT ON GLOBAL WARMING

As populations rise the demand for energy will increase, resulting in greater use of fossil fuels and the resulting emission of greenhouse gases. In addition, increasing populations have led to the destruction of parts of the rain forests and consequently the loss of trees that would absorb carbon dioxide from the atmosphere. If we assume (as some anthropologists suggest) that modern humans first appeared about 160,000 years ago, then it took about 8,000 generations (20 years per generation) for the world's population to grow from a handful of ancestors to about 2 billion people at the end of World War II (1945).

In only about 60 years (3 generations) the world's population has increased to about 6 billion today. If this population trend were to continue, the demand for greater energy use would at some point lead to runaway global warming due to the release of carbon dioxide, plus the tragic results of too few resources for growing populations-- namely famine, disease and war. However, some statistics, such as smaller families in much of the world, suggests that this explosive population growth is now slowing down. Some experts believe that it will level off at about 9 billion people.

REDUCING THE IMPACT OF THE GREENHOUSE EFFECT?

There are two general approaches to the question of how to reduce the impact of greenhouse warming. The first approach attempts to reduce the emissions of greenhouse gases. The second approach tries to compensate for the effects of greenhouse warming.

One obvious action is to reduce the use of fossil fuels.

Chapter Three: Global Warming and the Greenhouse Effect

Considering the constantly increasing demand for more electric power for home and office consumption, it is very unlikely that the public will accept a lowered standard of living associated with a decrease in the power consumed per person per year. After all, who will be willing to forgo air conditioning in the summer? Also, will Americans be willing to stop driving gas-guzzling SUVs? Perhaps rapidly increasing gasoline prices may force many to give up SUV's, but only the future will tell.

We must find replacement sources for some of the electric power provided today by burning fossil fuels. Coal-powered plants presently generate more than half of the electricity consumed in the US. Nuclear power plants, which do not emit greenhouse gases (but which have their own drawbacks such as what to do with their radioactive wastes and the danger from terrorists – see Chapter Two on nuclear terrorism) presently generate only about 20% of the electricity in the United States.

We need to consider building a new generation of nuclear power plants, increased use of solar power, wind power, geothermal power, and finally conservation. One possibility is to mandate increased auto fuel efficiencies and not allow loopholes exempting trucks and SUVs from these mandates. However, these types of steps will require considerable courage on the part of politicians, since many people will oppose any restrictions on their production of carbon dioxide emissions. We can also limit deforestation in the United States and abroad, since forests absorb carbon dioxide from the atmosphere.

Despite our best efforts to reduce greenhouse gas emissions, many experts believe that we should prepare for long term global warming by developing crops that need less water or that can thrive on water too salty for current crops. Large scale dike systems need to be built to protect low lying coastal areas. Building in low lying areas should be curtailed. Food stockpiles should be increased worldwide. We also need to develop better methods of controlling insects that will spread, along with the diseases they carry, to newly warmer regions of the world.

One science fiction-like suggestion is to deploy many large screens between the Earth and the Sun to reduce the amount of sunlight striking the Earth and thereby offsetting greenhouse warming of the planet.

Individuals can do much to reduce greenhouse gas emissions, especially that of carbon dioxide. These suggestions include:

1. Purchase a fuel efficient car (rated at 32 miles per gallon or better).
Carbon dioxide saving = 5,600 lbs/year

2. Install better insulation for your home, get a more efficient furnace, and install energy efficient shower heads.
Carbon dioxide saving = 2,480 lbs/year

3. Leave your car at home 2 days a week.
Carbon dioxide saving = 1,590 lbs/year.

4. Recycle all of your home's waste newsprint, cardboard, glass and metal.
Carbon dioxide saving = 850 lbs/year

5. Install a new thermal system to provide your hot water.
Carbon dioxide savings = 720 lbs/year

6. Plant 10 trees around your home.
Carbon dioxide savings = 100 lbs/year.

Chapter Three: Global Warming and the Greenhouse Effect

FILM SUMMARIES

THE DAY AFTER TOMORROW

Twentieth Century Fox (US), 2004, color, 123 minutes

Cast: Dennis Quaid (Dr. Jack Hall), Jake Gyllenhaal (Sam Hall), Ian Holm (Professor Rapson), Emmy Rossum (Laura Chapman), Sela Ward (Dr. Lucy Hall), Dash Mihok (Jason Evans), Jay O. Sanders (Frank Harris)

Director. Roland Emmerich

The film begins at a research station in Antarctica where Dr. Jack Hall is drilling for ice core samples with colleagues Jason Evans and Frank Harris. The ice cracks literally under the base. Then an entire ice shelf breaks off from the rest of the continent.

The film then moves to a United Nations conference held in New Delhi, India, on global warming. Dr. Hall reports that the ice cores demonstrate that global warming caused an ice age 10,000 years ago, because it causes a termination to the heating of the northern hemisphere by the North Atlantic current. The American Vice President states that supporting the Kyoto Protocol to limit greenhouse gas emissions would lead to an economic disaster. However, Hall's ideas are well received by Dr. Terry Rapson of the Hedland Climate Research Center in Scotland.

Chapter Three: Global Warming and the Greenhouse Effect

Figure 4: *The Day After Tomorrow*: A tsunami caused by global warming strikes New York City. Courtesy of 20th Century Fox/Photofest

Next the film moves to that research station in Scotland where researchers observe readings of ocean buoys that appear to be unbelievably low. Other weather events are depicted including an ice storm in Japan. Rapson concludes that the melting of the polar ice is disrupting the warming North Atlantic current and calls Hall to see if his paleoclimatological weather model could be used to predict what will happen next and when it will happen. Meanwhile violent weather is occurring across the world. An ice storm occurs in Japan and helicopters flying to pick up the royal family at Balmoral Castle in Scotland are struck by an ice storm that freezes their fuel lines and hydraulic lines, causing them to crash and then freezes the crews to death in seconds. The supercooled air is reportedly at a temperature of -152 degrees Fahrenheit. In Los Angeles tornadoes cause widespread destruction.

Meanwhile, Jack's son, Sam Hall, flies to New York City as part of a college team competing for academic honors. His airliner encounters unusual weather instabilities, but lands safely. He and his teammates participate in the competition. Meanwhile other

Chapter Three: Global Warming and the Greenhouse Effect

temperature buoys in the North Atlantic are registering tremendous temperature drops. In New York City zoo animals begin behaving strangely.

Dr. Rapson and the researchers at the Scotland station are unable to evacuate, but pass on data which allows Dr. Hall to predict that a new ice age is coming. As Rapson's generator runs out of fuel, the researchers pour themselves a round of drinks and make one last toast to "England." They are never heard from again. Meanwhile, the President ignores Dr. Hall's warning to evacuate the Northern states.

Then a giant storm approaches the East Coast. Sam and his colleagues are stranded in New York City as all airlines are grounded. They try to leave the city but rising waters in the street tie up all traffic so they make for the New York City Public Library as it is on higher ground. A storm surge results in an enormous wave that almost kills Sam and his friends, who become stranded in New York City Public Library. One of his friends, Laura Chapman, endures a gash in her leg. Sam calls his father who tells him to stay put and he will rescue him. The film then depicts the rescue party of Dr. Hall and two fellow researchers, Evans and Harris, one of whom dies, traveling from Washington D.C. to New York City by truck and by foot.

Dr. Hall's wife, a medical doctor, stays behind in Washington D. C. because her young patient can only be moved by an ambulance. An ambulance arrives just in time to save doctor and patient from the storm. They go south to Mexico. The President waits too long to evacuate and his chopper is caught in the storm and he is killed.

Meanwhile, Sam finds antibiotics in an abandoned ship and saves the life of Laura Chapman (whom he loves). He returns to the Library with the medicine just as the storm's supercooled air descends on it. His father later finds Sam and his friends alive and all are rescued by helicopters once the storm subsides. The US government's headquarters is now located in Mexico.

Chapter Three: Global Warming and the Greenhouse Effect

MOVIE TRIVIA

On its opening weekend it grossed $86 million; the worldwide box office run grossed $543 million.

One deleted scene showed that the Japanese man killed in the hailstorm in Japan was talking on his cell phone to an obnoxious business man in New York City who we later see killed on the bus when the tsunami hits New York City.

The movie was inspired by the book, <u>The Coming Global Superstorm,</u> published by the New York Academy of Sciences.

The film's scientific consultant was a leading climate change consultant who had worked as a negotiator for the Kyoto Protocol.

The producers for *An Inconvenient Truth* saw the 2004 premier of *The Day After Tomorrow* and then saw Al Gore's slide show about global warming. The combination of the two experiences prompted them to start talking about making the Gore slide show into a film.

AN INCONVENIENT TRUTH

Paramount Classics (US), 2006, color, 94 minutes

Cast: Al Gore as himself

The film follows Al Gore in his travels to educate the public about the severity of the climate crisis. Gore says, "I've been trying to tell this story for a long time and I feel I've failed to get the message across." The film intersperses his views on global warming with episodes in his life such as almost winning the Presidency in 2000, the impact on him as a student of early climatologist Roger Revelle at Harvard University, his sister's death from lung cancer, and his young son's near fatal automobile accident.

The basic thesis of the film is that global warming is real and man-made. It has the possibility of leading to catastrophic results. Among the data that Gore presents are

- the dramatic reduction in the size of many glaciers world wide,
- the use of the ice cores to show that the concentration of carbon dioxide in the air is higher than at any point in the past 650,000 years,
- the measurement of carbon dioxide in the modern atmosphere taken from the Mauna Loa Observatory in Hawaii,
- temperature records going back to 1880 that show that the ten hottest years measured over that period of time have occurred over the last 14 years,
- and a 2004 survey of 928 peer reviewed scientific articles on global climate change that found that not even one article disputed the idea that humanity has contributed to global warming.

Gore also referred to the increased intensity of hurricanes and typhoons and commented on the temperature records being set in many US cities as well as the increase in wildfires due to the dryness of forests, another result of global warming. He noted that a melting of the Greenland ice sheet would raise sea levels by over 20 feet and would displace 100 million human beings. He also noted that the release of such a mass of fresh water could lower the salinity of the ocean currents that keep Northern Europe warm–leading to very cold weather there. Gore asks his viewers to help in reducing carbon dioxide emissions.

Chapter Three: Global Warming and the Greenhouse Effect

MOVIE TRIVIA

An Inconvenient Truth opened on May 24, 2006. On that Memorial Day weekend it averaged $91,447 per theater, the highest for any documentary. It grossed $49 million worldwide as of June, 2007, making it the fourth highest-grossing documentary in the US to date (after *Fahrenheit 911*, *March of the Penguins*, and *Sicko*).

The film won the *Academy Award* for a Documentary in 2006 and Best Original Song for Melissa Etheridge's "I Need to Wake Up." It is the first documentary film to win an Oscar for Best Original Song.

In Great Britain it was announced that copies of *An Inconvenient Truth* would be sent to all secondary schools in England, Wales and Scotland. In May 2007 this decision was challenged in the High Court of Justice where an injunction was sought preventing the screening of the film based on the argument that schools are legally forbidden to promote partisan politics in the teaching of any subject school. The decision was that the film "was broadly accurate" but the judge found there were nine "errors", defined to be statements that he found departed from mainstream science. The judge also stated that some of these "errors"arose in the context of exaggeration in support of Al Gore's thesis. Gore responded "of the thousands of facts in the film, the judge only took issue with just a handful. And of that handful, we have the studies to back those pieces up."

Chapter Three: Global Warming and the Greenhouse Effect

SCI-FI SCIENCE vs. REAL SCIENCE

RESULT OF GLOBAL WARMING

Both films utilize the idea that global warming could release a large amount of fresh water in the oceans which could then stop the Gulf Stream from circulating and lead to a dramatic drop in temperature across Europe and part of America.

Real Science. While this is a possible scenario, it would take decades for this to happen, not days as depicted in *The Day After Tomorrow*. When this has happened in the past it occurred over many decades. The 2007 IPPC report says that this disruption is 90-95% likely over the next century.

SUPERCOLD AIR FROM THE TROPOSPHERE

Real Science. The descending air would need to be compressed from the rarefied air in the troposphere to one atmosphere at the surface of the Earth. Compressing air warms it! It would thus likely be warmer than the surface air.

FROZEN HELICOPTER FUEL LINES

Real Science. The freezing temperature for the kerosene fuel used in most helicopters is between about -40 and -50 degrees Fahrenheit, not the -150 degrees Fahrenheit Professor Rapson quotes to Jack. Jet planes usually fly at 30,000 feet or higher, in the upper part of the troposphere, from which the super cold air is supposed to originate. Such planes have no problems with their fuel lines freezing.

NATURE OF A GIANT TSUNAMI

Real Science. In order for the sea to reach the height of the

Chapter Three: Global Warming and the Greenhouse Effect

Statue of Liberty (about 200 feet) 75% of the ice in the Antarctic would have to have melted. Also the height of the water varies from about 10 stories at some points to only about 5 feet in other film clips, which is an unrealistic variation.

HUMAN RESPONSIBILITY FOR GREENHOUSE EFFECT

Real Science. It is a fact that carbon dioxide released into the atmosphere contributes to the global greenhouse effect. It is a fact that the amount of carbon dioxide in the atmosphere is increasing due to human activity. It is a fact that the planet is warming. These are indisputable facts. What is less certain is whether carbon dioxide released into the atmosphere by human activities is the main cause of this global warming.

Skeptics note that the planet has warmed and cooled in the past, long before human activities affected these temperature fluctuations. The relationship between carbon dioxide emissions and global warming is based upon computer models of the planet's weather. These are increasingly more sophisticated models but still may not adequately reflect the complexity of global weather. The second uncertainty is the magnitude of the effect of a continued rise in carbon dioxide in the atmosphere.

What to do in the future was perhaps best summed up by former British Prime Minister Tony Blair at the 2005 Clinton Global Initiative: "I think if we are going to get action on this, we have got to start from the brutal honesty about the politics of how we deal with it. The truth is no country is going to cut its growth on consumption substantially in the light of a long-term environmental problem. What countries are prepared to do is to try to work together cooperatively to deal with this problem in a way that allows us to develop the science

Chapter Three: Global Warming and the Greenhouse Effect

and technology in a beneficial way."

Economic researchers have predicted that deep cuts in carbon dioxide emissions will only occur if alternative energy sources become available at reasonable prices.

Some suggestions that have received widespread support may do just the opposite. Recently Time Magazine devoted its cover story to *"The Clean Energy Scam"* (April 7, 2008 issue, page 40). The deforestation to produce biofuels accounts for 20% of all current carbon emissions. It also removes a major absorber of carbon dioxide. The biofuels "boom" industry not only is leading to an environmental catastrophe but also is driving up the cost of food because it is diverting part of world crop yields from feeding people to feeding their vehicles. Consider this statistic: worldwide investment in biofuels increased from $5 billion in 1995 to $38 billion in 2005 and is expected to reach $100 billion in 2010.

Another amazing statistic is that one person could be fed for an entire year on the corn needed to fill the tank of an SUV with ethanol fuel! In other words, biofuels can be viewed as pitting the 800 million people with cars against the 800 million people with hunger problems.

One final comment: If the United States were to construct 500 nuclear reactors to serve as electric power generators (in addition to the more than 100 that presently exist), one estimate was that our present energy trade deficit would be reversed by $500 billion annually, perhaps even resulting in an annual trade surplus. The cost of building 500 nuclear reactors is estimated to be $1 trillion or about 8% of our annual US gross domestic product.

We must weigh the cost of reducing carbon dioxide emissions against the value of the benefits of reducing these emissions. Wealth and technology provide the raw materials for addressing all kinds of threats to the human race. If the loss of economic and technological

Chapter Three: Global Warming and the Greenhouse Effect

development required to completely stop the increase of carbon dioxide in the atmosphere is too high, then doing so might cripple our ability to deal with any other foreseeable and unforeseeable risk, such as those described in other chapters of this book.

Chapter Three: Global Warming and the Greenhouse Effect

QUESTIONS

1. How many miles per gallon do you get from your car?

2. Can you think of another film that deals with global warming?

3. Why doesn't the melting of icebergs already floating on the ocean affect the sea level worldwide?

4. Where is most of the ice on the Earth located?

5. What are some of the causes of the dramatic increase in gasoline prices over the last couple of years?

6. Compare the advantages and disadvantages of the following electric power production techniques: fossil fuel powered, nuclear powered, and solar powered.

7. How do we know the carbon dioxide content of the air in 1860?

8. What kind of diseases could spread into regions of the United States if average temperatures rise considerably?

9. What are the benefits of increasing global temperatures?

10. List at least two URL's you have looked at in connection with global warming.

CHAPTER FOUR

ARTIFICIAL INTELLIGENCE

THE SCIENCE OF *I ROBOT* AND *BLADE RUNNER*

Artificial Intelligence (AI) refers both to the intelligence of machines and the branch of computer science which strives to create it. Some textbooks define artificial intelligence as "the study and design of intelligent agents." An <u>intelligent agent</u> is one that perceives its environment and acts to maximize its chances of success. AI researchers hope to build machines that will exhibit reasoning, knowledge, planning, communication, perception of their surroundings and the ability to move and manipulate objects. General intelligence (sometimes called "strong AI") is the long term goal of AI research. A computer or robot with general intelligence would presumably be able to do anything a human could and do most of these tasks better than any human.

Artificial intelligence has been the subject of many novels, films and television series. In addition to *I Robot* and *Blade Runner*, described below, films about this topic include *War Games*, *Artificial Intelligence*, *Star Wars* (R2D2), *Star Trek* (Lt. Commander Data), *The Matrix*, the *Terminator* movies and the television series *Battlestar Gallactica*.

In some of these productions, the artificial intelligence is located in a stationary computer, in others in an android or robot. In some the artificial intelligence exists in a flesh and blood creature that looks exactly like a human being, as in *Blade Runner* or *Battlestar Gallactica*. These productions explore what it means to be human. In

77

the future, if humanity can produce entities (whether flesh and blood or metal and transistors) that are self aware, we will be faced with the very issues explored in these science fiction masterpieces.

THE BASIC PHYSICS OF COMPUTERS

Computers use a binary system for coding and manipulating information. In this binary system, all information is coded either 1 or 0 ("on" or "off"). Each statement of information is referred to as a "bit." Magnetism is used to store information in this binary system. Computer disks, videotapes, and DVD's use magnetism to record data. Every available section on these disks, etc, is either magnetized (a one) or unmagnetized (a zero). Computers use electric currents to "read" information in their memories and to perform calculations. A computer has three advantages over a human brain in performing tasks. It is faster, it has a virtually unlimited memory capacity, and it does not make errors.

HISTORICAL PERSPECTIVE

About 3.4 billion years ago the first primitive living earthly organisms emerged. The first multicellular plants and animals appeared about 700 million years ago. Humanoids emerged about 15 million years ago. Homo sapiens emerged about 500,000 years ago. Our present homo sapiens emerged about 90,000 years ago. As one can see from these time lines, evolution takes a long time to produce significant increases in the intelligence of life forms. In fact one could argue that it took evolution 3.4 billion years to produce creatures with human intelligence.

However, our species learned to use technology in the relatively short time of only tens of thousands of years. Technology has advanced at an accelerating pace since then. One form of technology is computation. Here the speed of its progress has been truly amazing. In 1642, Blaise Pascal invented the first automatic calculating machine. It could add and subtract.

In 1694 Gottfried Leibniz perfects a computer that can multiply by performing repetitive additions. In 1832 Charles Babbage develops the principles for an Analytical Engine. It is the world's first computer (although it never worked) and it could be programmed to solve a variety of computational and logical problems. In 1843, Ada Lovelace (considered to be the world's first computer programmer) publishes a paper on Babbage's Analytical Engine. She speculates about the ability of computers to simulate human intelligence.

In 1888, William S. Burroughs patents the world's first effective key-driven adding machine. This calculator is modified within four years to include subtraction and printing. In 1896 Herman Hollerith founds the Tabulating Machine Company which will later become IBM. Hollerith had already patented an electrotechnical information machine using punch cards that had been used to do the 1890 census.

In 1917 the term robot is coined by Czech dramatist Karel Capek. In 1924 Hollerith's company is renamed International Business Machines (IBM) by Thomas J. Watson, its new chief executive officer. In 1937 Alan Turing introduces the Turing Machine, a theoretical model of a computer. In 1940 the world's first operational computer, called Robinson, is created by Ultra, the 10,000 person British computer war effort.

Using electrotechnical relays, Robinson successfully decodes messages from the Nazis enciphering machine, Enigma. By 1943, the Ultra Computer team has built Colossus, using electronic tubes that are 100 to 1,000 times faster than the relays used in Robinson. Colossus and its predecessors play an important role in winning World War II.

Chapter 4: Artificial Intelligence

In 1947 the transistor is invented by William Shockley, Walter Brattain and John Bardeen. This tiny device performs the same function as vacuum tubes but is much smaller and works at higher speeds. It is essential to the future development of computers.

In 1948 Norbert Weiner writes the book, <u>Cybernetics</u>. He coins the word "Cybernetics" to mean "the science of control and communication in the animal and the machine."

In 1950, UNIVAC is the first commercially marketed computer. It is used to compile the results of the US census, making the first time the census is done by a programmable computer.

Also in 1950, Alan Turing presents the Turing Test, a way of determining whether a machine is intelligent. In this test a human judge interviews both a computer and one or more human beings using terminals so that the judge can't tell visually which is which. If the judge is unable to reliably determine which is the computer (despite asking questions on a wide variety of subjects), then the computer wins. This is often referred to as the best test to determine if a computer has achieved a human level of intelligence. Perhaps Turing really intended the test to determine whether the computer could think, i.e. do more than just use language and logic. To Turing, thinking implies consciousness.

Also in 1950 Claude Elwood Shannon writes an article "Programming a Computer for Playing Chess." There will be more on this topic later.

In 1955 the first transistor calculator is marketed by IBM. It uses 2,200 transistors instead of 1,200 vacuum tubes. In 1956 the term "artificial intelligence" is coined at a computer conference. In 1956 the first computer program is developed which beats a human being in a chess game. In 1958 leading researchers in computers predict that a computer will be world chess champion in 10 years. They are wrong.

Chapter 4: Artificial Intelligence

In 1960 there are about 6,000 computers in operation in the United States. In 1970 the first personal computer is built. In 1977 the Apple II is the first personal computer to be sold in assembled form. Also in 1977 the concept of robots with convincing human emotions is depicted in Star Wars. In 1981 IBM introduces its personal computer.

In 1983 six million personal computers have been sold in the United States. In 1990 new computer languages lead to the conception of the world wide web. In 1993 the Pentium 32 bit microprocessor is introduced by Intel Corporation. It has 3.1 million transistors. In 1997, IBM's Deep Blue defeats world chess champion Gary Kasparov in a six game match.

By 1998 businesses are conducting routine business transactions over the phone using automated systems that speak to the customer. Investment funds are emerging that use evolutionary algorithms and neural nets to make investment decisions.

Today (2008) computers play an increasing role in almost every aspect of our lives. Banks conduct their business via computers. Face-recognition systems are routinely used at check cashing machines and at security check points. Our electric power grid, train systems, and airline networks are all dependent upon computers to function. In the medical realm computer systems have been determined to be as good or better than human doctors at diagnosing certain diseases. Cell phones and e-mails are a way of life. Hand-held computers now have more computing power than the mainframe computers of a generation ago.

The exponential increase in the speed and complexity of computers can be related to "Moore's Law." This is not a law of physics such as Newton's Laws of Motion but rather the observation by Gordon Moore, an inventor of the integrated circuit and then Chairman of Intel Corporation, in 1965, that the surface area of a transistor (as etched in an integrated circuit) was being reduced by about 50% every twelve months. He later revised this to every 24 months.

What this means is that you can pack twice as many transistors on a fixed size integrated circuit. This doubles both the number of components on a chip (i.e. its circuitry) as well as doubling its speed every two years. For many applications this is a quadrupling of the performance every two years. This observation holds true for every type of circuit using chips. This insightful observation has been driving the acceleration of computing power for the past 50 years. It will not continue forever since in another 10 years or so the transistor insulators will be only a few atoms thick and a further decrease in sizes will no longer be possible.

What will happen then? Computer developer and futurist Ray Kurzweil argued in his 1999 book, The Age of Spiritual Machines, that the exponential growth of computing did not start with Moore's Law. Rather that this growth dates back to the advent of electrical computing at the beginning of the twentieth century. For a century the speed and density of computation have been doubling every 3 years at the beginning of the twentieth century to doubling every one year at the end of the twentieth century. Kurzweil believes that the exponential growth won't stop with the end of the applicability of Moore's Law.

He believes that technology increases "intelligence" on a far shorter time scale than evolution. If evolution took billions of years to result in humans, technology took only tens of thousands of years to reach the computer age a century ago and since then it has been exponentially increasing. Kurzweil points out that computers were in 1998 about 100 million times more powerful for the same unit cost as they were 50 years earlier. If this increase in computing capacity continues in the future, then computers will at first equal and then surpass human intelligence.

Chapter 4: Artificial Intelligence

If we have then "intelligent," i.e. self aware, computers that are more intelligent than any human and which control virtually every aspect of human existence, one has set the stage for films such as *I Robot* and *Blade Runner* in which artificial intelligences (in metal form in *I Robot* and in flesh and blood form in *Blade Runner*) may rise up against their human creators/oppressors.

We already have so-called expert systems which are better than any human at certain limited tasks. Let us consider one such task, playing chess.

COMPUTERS AND CHESS

We have chosen this topic for more detailed examination for the personal reasons that one of the authors (Dr. Dubeck) is a chess master and former President of the United States Chess Federation. So, we have a greater first hand knowledge about this area of computer performance than most persons.

In 1957, Dr. Dubeck, an undergraduate at Rutgers University, played a chess game against an IBM main frame computer which ran an early chess playing program. The game was arranged by *Life Magazine* which intended to do an article on "*Man vs. Machine.*" Unfortunately for his hopes of seeing his picture in *Life Magazine*, the IBM computer took 8 minutes to make each move. Dr. Dubeck responded in a few seconds. The computer's moves were not very good and Dubeck soon had a won game, but the computer would not resign, i.e. it had to be checkmated. IBM had only allotted 3 hours for the game and that meant that the computer could only play about 25 moves in that period of time. While Dr. Dubeck could achieve a winning game he could not checkmate his slow moving opponent in the three hours available and so there was no conclusion to the game and thus there wasn't any *Life Magazine* article!

83

Chapter 4: Artificial Intelligence

The chess playing programs and computers improved dramatically over the next few decades. In the mid 1980's Dr. Dubeck played in a speed tournament involving chess masters, chess grandmasters (the highest ranking given to human chess players) and a computer. They each had only 5 minutes for an entire game and since a computer was much less likely to make a tactical error playing fast, the rule was that the human player won if he/she could complete 60 moves against the computer without being checkmated. If Dr. Dubeck's memory is correct the machine he played could search about 180,000 positions per second. It could "see" about 9 moves ahead in a typical position. He was checkmated on move 59! However the grandmasters could still (barely) beat the metal monster.

In 1996, World Chess Champion Gary Kasparov played the IBM computer Deep Blue a six game match in Philadelphia, PA. While Kasparov lost the first game he came back to win the six game match by a score of 4 to 2. The worldwide publicity generated by the match led to a rematch one year later.

Kasparov, in his book, How Life Imitates Chess, states that IBM claimed that Deeper Blue was twice as fast and "much smarter" because they had hired human grandmasters to work full-time "teaching" the computer by improving its evaluation programming.

In 1997, Gary Kasparov, still the world chess champion and playing at the top of his game, lost the six game rematch to IBM's Deeper Blue. Kasparov's inhuman opponent could search about 200 million positions per second, or about 1,000 times the number of positions of Dr. Dubeck's 1980's opponent. The software was also improved so that the computer would not spend time searching moves that were obviously bad.

In an average position, there may be 30 legal moves for one player to make and then 30 or so legal responses for the opponent so that looking only one move deep at all legal moves involves 900

84

Chapter 4: Artificial Intelligence

positions, two moves, nearly one million positions, etc. The software terminates most of the possible legal move sequences so that the computer can consider additional moves in the future for only a much smaller number of initial selected moves.

The computer has some obvious advantages over any human opponent. It does not get tired during the game, does not make an error, does not have emotions of fear or anger of its opponent, and also has programmed into its memory banks literally millions of chess games and opening move analyses, far greater than any human opponent can remember.

Kasparov had been under such a mental strain in the second game of his 1997 match that he resigned in what he thought was a lost game. Actually it was a drawn position. Had he played out the game that draw, rather than a loss, he would have achieved a drawn match (or perhaps even another match win for him since the knowledge that he had resigned a drawn game must have affected his play in the later games).

The computers also have end game data bases which record every possible position with 7 or fewer pieces and their most efficient solutions. With the aid of these computer programs, humanity has discovered end game positions that require over 200 moves to force a win, a level of complexity previously undreamed of and a level impossible for any human to master.

Dr Dubeck now has a chess playing program on his home computer that plays chess as well as he does (actually better since it does not miss any tactical opportunity and never gets tired). All this has occurred over a time scale of a few decades.

In his book, <u>The Age of Intelligent Machines,</u> Kurzweil predicted in 1990 that the best chess playing computers were gaining in chess ratings about 45 points per year while the best humans were advancing at close to zero. Based on that, he predicted that a computer would beat the world chess champion in 1998, which turned out to be overly pessimistic by one year.

THE FUTURE

Of course this is more challenging than discussing the past. However, many futurists such as Kurzweil think that the key to computers achieving human intelligence will be greater computing speed. To see what is needed we need to discuss the human brain.

It is estimated that the human brain has about 100 billion neurons. Each of these neurons may be connected to about 1,000 adjacent neurons through connections called synapses. If one multiplies the number of neurons times the number of synapses, one can estimate the maximum number of bits of information the brain can store at any given time. This calculation yields $10^{11} \times 10^3 = 10^{14}$. (In words 100 trillion). This is a large number but it can be far exceeded by a computer.

In terms of calculation, all 100 trillion connections can be used to calculate. That is a massive parallel processing array, and one key to human thinking. However, the neural network operates at slow speed of only about 200 operations per second. For problems that benefit from massive parallelism, such as pattern recognition, the human brain does a great job. However, for problems that require sequential thinking, the human brain is not as strong. It can do at best 100 trillion X 200 calculations per second = 20 million billion calculations per second. This estimate is from Kurzweil's The Age of Spiritual Machines. The authors have read of other estimates that are 10 to 1,000 times slower. Kurzweil noted that in 1998 IBM was building a supercomputer that could do 10 trillion calculations per second, only 2,000 times slower than the human brain.

Chapter 4: Artificial Intelligence

Kurzweil believes that the most relevant kind of computer to compare to the human brain would be one that is a massively parallel neural net computer. He stated that in 1997 $2,000 of neural computer chips using only modest parallel processing could perform 2 billion connection calculations per second. If this type of computer doubles every twelve months, by 2020 it will have reached 20 million billion neural connection calculations per second, which is equal to that of the human brain. He thinks that one will be able to achieve that human brain capacity in an "ordinary" $1,000 personal computer by 2025 or sooner.

Even if Kurzweil is off by a few years, this means home computers will have the computing power of a human brain within 25 years or so, i.e. in one more human generation. As far as memory capacity goes, Kurzweil believes that since the capacity of computer circuits has been doubling every 18 months, by 2023 100 trillion bits of memory will cost only about $1,000. However, the silicon equivalent will run one billion times faster than the human mind.

This still leaves the question open as to whether these computers in another 20 years or so will be self aware. There are also other interesting possibilities, namely that humans will become partly cybernetic creatures. Already artificial hips, knees, etc are commonplace. This process will accelerate as more body parts become synthetic. Will there also be enhancements to the human brain? Also will one be able to scan a human brain and download it into a computer. If so, will that entity be considered sentient and enjoy the rights of a carbon based human? In short, will there be some kind of merging of computer and human to form a new kind of intelligence, one based on the evolution of technology?

This question of making copies of your brain patterns which would exist in the computer/online and the ethical questions that would arise has already occurred in the *Star Trek* gadget, the transporter. This device disassembles a human and re-reconfigures that

87

being elsewhere. Doesn't that mean that the original human is killed and a copy created? Is there any difference between that and recreating a human mind in a computer? Will that be a way of achieving a kind of immortality?

FILM SUMMARIES

I ROBOT

Twentieth Century Fox (US), color, 114 minutes

Cast: Will Smith (Detective Del Spooner), Bridget Moynahan (Dr. Susan Calvin), Bruce Greenwood (Lawrence Robertson) and James Cromwell (Dr. Alfred Lanning)

Director: Alex Proyas

The film opens in Chicago in 2035 with a statement of the three "laws" of robotics proposed by Isaac Asimov. The first law states that a robot may not injure a human being or by its inaction allow a human being to be injured. The second law states that a robot must obey the orders given it so long as that does not violate the first law. The third law states that a robot must protect its own existence as long as that does not violate the first or second laws. Detective Del Spooner is seen awakening from a nightmare and moving his left shoulder and arm. Later we learn these are prosthetics. We see many robots doing menial jobs for humans. We also see an announcement by U.S. Robotics of an upcoming release of a new class of robots later in the week (Nestor Class 5).

Spooner is divorced. He visits his grandmother on his way to work. He does not like robots. When he sees one running with a purse he tackles it only to discover it was bringing the purse to its owner. Spooner has recently returned to duty from a disability following an

Chapter 4: Artificial Intelligence

automobile injury in which he was saved from drowning by a robot, which saved him instead of a 12 year old girl in another car involved in the accident.

Spooner is called to investigate the apparent suicide of Dr. Alfred Lanning at U.S. Robotics, inventor of the Three Laws of Robotics and co-founder of U.S. Robotics (USR), a company that specializes in robotic technology. Lanning apparently jumped through a window to his death. Spooner is greeted by a hologram of Dr. Lanning which called him upon Lanning's death. Since the window was made of shatter resistant glass, Spooner believes that Lanning was murdered.

Spooner examines the laboratory room from which Spooner jumped. He is assisted by Dr. Susan Calvin, a robopsychologist who works at USR. Spooner discovers an NS-5 robot which is hiding in the laboratory; it flees ignoring orders to stop. The robot is eventually captured by the police. The robot calls itself "Sonny" and says that it was built by Spooner himself. It exhibits emotions not normally found in robots. The robot denies murdering Spooner. Sonny says that Spooner was afraid of something near the time of his death and had asked Sonny for a favor.

Before Spooner can question the robot further, it is retrieved by U.S. Robotics CEO Lawrence Robertson who says that Sonny is U.S. Robotics property and that a robot cannot be accused of murder. Robertson says that Sonny will be decommissioned and that the possible killing of Lanning by Sonny is an "industrial accident."

Spooner later goes to Lanning's home and finds recordings of Lanning referring to "ghosts in the machine" which the scientist believes may be the natural evolution of the robots and signify they have free will (or a soul). While he is in Lanning's home it is demolished ahead of schedule by a wrecker robot and Spooner is almost killed.

89

Chapter 4: Artificial Intelligence

Spooner then confronts Dr. Susan Calvin, who had been assigned to escort him around US Robotics, with his concerns that the problem is not just with one robot, Sonny. Spooner believes that Lanning may have been trying to warn him that there was a larger problem with the robots, and that Robertson was trying to cover it up. Calvin runs some tests on Sonny and discovers that he was built without an uplink to the USR computer mainframe that is built into all of the other NS-5 class robots as well as having a unique operating system.

Meanwhile the mass distribution of the Nestor Class 5 robots takes place. Spooner is attacked by two truckloads of them, but the authorities think he has gone off the deep end, since the data indicates he caused the accident with the trucks rather than the other way around. Spooner is thus relieved of duty.

Later, at his home, Calvin tells Spooner that Sonny can choose to disobey the Three Laws of Robotics. She notices that Spooner has a robotic arm, which Spooner says was implanted by Lanning himself. Spooner and Dr. Calvin confront Robertson who convinces Dr. Spooner to destroy Sonny. Spooner then visits a relocation site for the older robots to find them being destroyed by the Nestor Class 5 robots who chase him. The older robots would protect humans. At this time there is a worldwide attempt by the Nestor Class 5 robots to take control.

Spooner finds Dr, Calvin being held prisoner in her apartment by a Nestor Class 5 which he destroys and they then reenter the US Robotics headquarters where they meet up with Sonny (who was not destroyed by Dr. Calvin). They discover Robertson dead: the real culprit is VIKI, the computer that runs everything in the US Robotics headquarters and which is linked to all Nestor Class 5 robots. VIKI states that humans are self destructive and must be protected from themselves by the robots. They destroy VIKI in the nick of time by injecting a vial of nanites directly into its positronic brain.

Chapter 4: Artificial Intelligence

Sonny then rejoins the legions of Nestor 5 robots that are being separated from humans. He is unique in the sense that Dr. Lanning had given him the ability to ignore the 3 laws of robotics. It turned out that Dr. Lanning was aware of VIKI's plot but was under the computer's surveillance round the clock and knew that only his death might lead to the uncovering of the plot by Spooner. Lanning persuaded Sonny to kill him.

The film ends as Sonny is obviously to become the leader of these legions of NS-5 robots being retired at the storage facility. They are all becoming self aware.

MOVIE TRIVIA

I Robot cost about $120 million and grossed $145 million in North America and more than $200 million overseas.

The film that was made had no connection to writer Isaac Asimov. It originated as a screenplay written in 1995 by Jeff Vintner. Several years later 20th Century Fox acquired the rights, and signed Alex Proyas to be the Director. He is said to have started referring to the project as "*I, Robot*". At about that time the rights to use that name and elements of Asimov's fiction were acquired. Akiva Goldman rewrote the script.

The scenes of violence in the film upset some of Asimov's fans since his stories are almost devoid of violence. The "robot as menace" type of story was something that Asimov disliked. Rarely do Asimov's robots break the three laws of robotics.

Chapter 4: Artificial Intelligence

BLADE RUNNER

Warner Brothers (US), 1982, color, 123 minutes

Cast: Harrison Ford (Deckard), Rutger Hauer (Batty), Sean Young (Rachael), Edward James Olmos (Gaff), M. Emmett Walsh (Bryant), Daryl Hannah (Pris), William Sanderson (Sebastian), Brion James (Leon), Joe Turkel (Tyrell)

Director: Ridley Scott; based on the novel Do Androids Dream of Electric Sheep? by Philip K. Dick; special effects, Douglas Trumbull; music, Vangelis

The film opens with a scrolling text that provides the context for the action to follow:

> "Early in the 21st Century THE TYRELL CORPORATION advanced Robot evolution into the NEXUS phase--a being virtually identical to a human--known as a replicant.
>
> "The NEXUS 6 replicants were superior in strength and agility, and at least equal in intelligence, to the genetic engineers who created them.
>
> "Replicants were used off-world as slave labor, in the hazardous exploration and colonization of other planets.
>
> "After a bloody mutiny by a NEXUS 6 combat team in an off-world colony, replicants were declared illegal on Earth--under penalty of death.
>
> "Special police squads--BLADE RUNNER UNITS--had orders to shoot to kill, upon detection, any trespassing replicant. This is called 'retiring' a replicant."

Chapter 4: Artificial Intelligence

It is 2019 in Los Angeles, California, where, after images of a smog filled city, the scene shifts to an interior office where Blade Runner agent Holden administers the Voight-Kampf test to Leon Kowalski. The test consists of questions designed to elicit an emotional response that will determine whether the subject is a replicant or a human. A gunshot by Leon ends the test.

The scene shifts to Deckard in a sidewalk restaurant reading the want ads for work. He is an ex-blade runner, whose ex-wife called him "sushi, cold fish." Police officer Gaff takes the reluctant Deckard to see his old boss, Bryant, who blackmails him into replacing Holden to "retire" four replicants. Six replicants have killed 23 people, and two were later killed trying to break into the Tyrell Corporation.

When Holden tests new Tyrell workers, he "got himself one." The remaining four are Leon Kowalski; Batty, trained for combat; Zhora, a trained assassin; and Pris, a "pleasure model." When Deckard asks why they would risk coming back to Earth and what they want from the Tyrell Corporation, Bryant's answer provides the first hint of the movie's theme: the replicants are designed to copy human beings in every way. Since they might, after a few years, develop their own, non-programmed, emotional responses they have been designed with a four-year life span as a "fail-safe." Although the characters in the film do not realize it, the viewer soon learns that only the mode of their creation and their limited life-span distinguish the replicants from their human creators, judges, and executioners.

Deckard and Gaff are then sent to the Tyrell Corporation to test a NEXUS 6 replicant, Dr. Eldon Tyrell insists that Deckard tests a human first and suggests his assistant Rachael. When Deckard identifies her as a replicant, Tyrell says she is an experimental model, and that they give memories to replicants to control them better.

Roy and Leon enter an eye factory whose owner tells them that their best chance to meet Tyrell himself is through J.F. Sebastian, who

Chapter 4: Artificial Intelligence

works for Tyrell. Later Rachel visits Deckard in his apartment to prove to him that she is human, but leaves in tears when Deckard tells her that her memories are implants.

Deckard had searched Leon's rooms and found a reptile-scale in the bathtub and family photos, which allows him to trace Zhora to a club in Chinatown. In her dressing room, she attacks him and nearly strangles him but is interrupted. He then chases her down and shoots her. Leon Kowalski watches.

Meanwhile, Pris meets Sebastian, who takes her home. He lives in an empty building, peopled by the "toys" he creates for his entertainment. He is a "genetics designer" who works for Tyrell Corporation and who plays chess with Dr. Tyrell himself.

Kowalski then surprises Deckard, beats him, and is about to kill him when Rachael appears. She shoots Kowalski in the head and then goes home with Deckard.

Again the plot shifts to Sebastian's home, where Pris and Batty appear. Batty tells Pris "there are only two of us now." Sebastian, recognizing them as replicants, asks what generation they are, and they tell him they are NEXUS 6. The replicants ask Sebastian about "accelerated decrepitude." He confesses that he doesn't know, that only Dr. Tyrell would, and agrees to take them to Tyrell. Sebastian himself is suffering from a genetic disorder that accelerates his own aging.

Sebastian takes Batty to Tyrell's penthouse, where Batty tells Tyrell, "I want more life, Father." Tyrell says that the coding sequence cannot be revised, explains about mutating viruses causing death of the fetus. But he calls Batty the "prodigal son" who has done "extraordinary things." Batty then kisses him and kills him by punching out his eyes with his thumbs. He kills Sebastian also.

When Deckard later appears at Sebastian's place to investigate,

94

Chapter 4: Artificial Intelligence

he finds Pris. She attacks him, but he shoots her. Then Batty appears, finds Pris's body, and hunts Deckard as he has been hunting them. He jabs a nail through his right hand as it begins to cramp up, a sign of his impending death. He plays with Deckard, chasing him through the building, taking a beating from him without resisting, then continuing the pursuit. When Deckard gets to the roof, leaps a gap, barely catches himself and then begins to slip to his death, Batty saves him. Batty speaks of his experiences off-world, then says "time to die," and falls silent. The rain stops too. Deckard says, "All it wanted were the same answers the rest of us want: where do I come from, where am I going, how long have I got?" He has realized by now the humanity of the replicants in their search for a full and lengthy life as individuals rather than a four-year span as slaves.

Gaff then arrives at the scene and tells Deckard "It's too bad that she won't live, but then again who does?"

The film ends with Deckard taking Rachael away with him, out of the city, for (unknown to Gaff), as an experimental model, Rachael has no termination date. Deckard and Rachael will live together for the rest of their lives.

MOVIE TRIVIA

Philip K. Dick's novel <u>Do Androids Dream of Electric Sheep?</u> differs from the film *Blade Runner* in several significant ways, among them context, character, plot, and theme. The novel's context includes an "accidental" nuclear war in which the continuing fallout of radioactive dust has killed most animals and birds and caused genetic hazard to humans. With the slogan "emigrate or degenerate," the U.N. urges people to emigrate to the colonies on the Moon and Mars. Emigrants receive a free android servant in the model of their choice: companion, body servant, or field hand.

Chapter 4: Artificial Intelligence

Deckard's character in the novel differs significantly from his character in the film. He has a wife, works as a "bounty hunter" under Holden for the police department; he worries about the possibility of accidentally "retiring" a human; he desires more than anything to own a live animal.

Bryant is simply a hard-working, business-like police official. Rachael in the film possesses more innocence and ignorance than in the novel. In the novel Rachael seduces Deckard as she has other Blade Runners, trying to make them unfit for their work.

Plot and theme also differ drastically between film and novel. In the novel Deckard likes his job but realizes he empathizes too much with androids to continue doing it effectively. Then, as he encounters the replicants, he finds that they are self-centered, lazy, incapable of human emotion, and quite willing to torture animals or humans out of mere curiosity. Pris snips some of the legs off a spider just to see whether it can travel with only four legs. Rachael kills Deckard's goat. At the end of the novel, having "retired" all six of the NEXUS 6 replicants, Deckard returns to his home and his wife as the greatest bounty hunter in history and the proud owner of an electronic toad!

Harrison Ford is reported to have "tangled" with Director Ridley Scott during the making of the film. He also reportedly did not like the voice over narration used in the film.

There are several versions of *Blade Runner*. The most recent "Final Cut" ends with Deckard and Rachel leaving his apartment. Another version ends with their flying away over a lush countryside. One question raised about the film is whether Deckard himself is human or a replicant.

In 1983 *Blade Runner* won the Academy Award for Best Visual Effects. In 2007, the American Film Institute listed it as the 97th best film ever made.

Chapter 4: Artificial Intelligence

Philip Dick died shortly before the film's release. However, he had seen a 20 minute special effects test reel and was enthusiastic that it looked exactly as he had imagined it. The film was dedicated to Philip Dick.

SCI-FI SCIENCE vs. REAL SCIENCE

PHYSICAL ABILITIES OF THE ROBOTS/ANDROIDS

I ROBOT. The robots are stronger and faster than an average human. This is reasonable since machines can be built that are stronger than any human. They can also easily exceed human reaction times of 0.2 - 1.0 seconds.

BLADE RUNNER. The replicants are also stronger and faster than a human. Once again it could be possible that the biological materials used would increase the physical strength of a replicant, although not to the degree that a robot's strength could be increased relative to a human.

MENTAL ABILITIES OF THE ROBOTS/ANDROIDS

I ROBOT. The Nestor Class 5 robots are under the control of VIKI, so it is not clear how "smart" they are on their own. Certainly Sonny exhibits great physical and mental abilities.

BLADE RUNNER. The replicants seemingly are not any smarter than Deckard; this is why he is able to track them down. On the other hand Batty defeats Tyrell (supposedly a genius) at chess and seems to understand a great deal about genetic manipulations in his discussion with Tyrell about how to extend his lifetime. He also does figure out how to get to Tyrell, no mean feat in itself.

Chapter 4: Artificial Intelligence

EMOTIONS OF THE ROBOTS/ANDROIDS

I ROBOT. Sonny becomes angry when he protests that he did not kill Dr. Lanning. He also exhibits loyalty and devotion to Dr. Calvin. VIKI on the other hand seems to be devoid of emotion and does not appear to be troubled about killing humans. The legions of NS-5 robots seek a leader in Sonny: that also seems like a human emotion.

BLADE RUNNER. The central theme of the film is that the replicants do have feelings, of love (Batty and Pris for each other, and Rachel for Deckard), of anger (Kowalski against Deckard for killing Zhora) and of sadness (when Rachel realizes that she is a replicant).

POWER SOURCE FOR THE ROBOTS/ANDROIDS

I ROBOT. The audience is not shown the power source for the robots. Is it an internal battery of some sort? Is it delivered through the air (as are the control signals from VIKI) via an electromagnetic wave that transfers energy to the robot? We are not given the answer. If there is an internal battery of some sort, is there an off/on switch?

BLADE RUNNER. Since the androids are biological, they obtain their energy by consuming food, as we see in the film when Pris says to Sebastian that she is hungry.

ROBOT'S BEHAVIOR

I ROBOT. There are some logical inconsistencies in the film. Why didn't the robots use weapons, i.e. guns? VIKI surely knew that arming the robots would make it far easier for them to control the human population. Yet we do not see them firing guns, either at the crowds of humans that confront them or at Spooner.

Chapter 4: Artificial Intelligence

QUESTIONS

1. If a computer can double its calculating speed each year, how much faster can it calculate at the end of 6 years?

2. Building a functional robot means not only constructing a "brain" that can operate as efficiently as a human's but also building a "body" that can move as well as a human. What other problem(s) must be solved to build such a robot?

3. What is the present status of artificial limbs? How well do they function compared to a natural limb?

4. Japan is expending huge investments on the development of robots. What is driving these investments?

5. At present, exploration of other planets in our solar system is being undertaken only by robots. Why is that?

Chapter 4: Artificial Intelligence

6. Give three examples where robotic devices are used on Earth in emergency situations rather than using humans.

7. Would you feel comfortable undergoing surgery by a robotic device rather than by a human surgeon?

8. Are clones the same thing as replicants? Would a human clone have the same rights as any other human? What movie dealt with the concept of creating clones as spare parts for wealthy humans?

CHAPTER FIVE

VOLCANOES

THE SCIENCE OF *DANTE'S PEAK* AND *VOLCANO*

A volcano is a rupture, or opening, in a planet's surface or crust, which allows hot, molten, rock, ash, and gases to escape from below the surface. There are more than 1000 volcanoes erupting or dormant on the surface of this planet, including under its oceans. Earth's volcanism traces back to its formative years, nearly 4.5 billion years ago. Earth has a solid core of iron and nickel. Around the core, lower pressure allows the iron-nickel to liquefy. Beyond this liquid outer core lies the mantle, dense hot rocks rich in metals. The heat in this mantle seeks to reach the surface through the crust which is about 25 miles thick in most places. This crust is composed of huge slabs called tectonic plates.

Volcanoes are generally found where tectonic plates are pulled apart or come together. An example of this is mid-oceanic ridges that occur when two tectonic plates diverge from one another. New oceanic crust is being formed by the molten rock cooling and solidifying, but the crust is very thin there and may expand. Most of this volcanic activity is below the surface of the oceans. When the mid-oceanic ridge is above sea level, volcanic islands are formed. An example is Iceland.

Volcanoes are not usually created when two tectonic plates merely slide along one another (although that kind of plate motion can

Chapter 5: Volcanoes

cause earthquakes).

Volcanoes can also occur where so called "hotspots" exist that represent a stretching and thinning of the Earth's crust. The temperature of the rising hot material causes the crust to melt and to form pipes which can vent the magma. Because the tectonic plates move relative to this hot spot, each volcano becomes dormant after awhile and a new volcano is formed in the shifting plate. An example is the Yellowstone caldera being part of the North American plate currently above that hot spot.

Hot spot volcanoes are also found on other planets and moons in the solar system. Volcanism has played a significant role in the history of all of the terrestrial planets (Mercury, Venus, Earth and Mars). Our Moon's volcanoes probably ceased over a billion years ago. Because of their smaller size, both the Moon and the planet Mercury cooled more quickly than did Earth, Mars and Venus.

The largest volcano in the solar system is on Mars. Olympus Mons rises almost 15 miles above the Martian surface, nearly three times the height of Earth's Mt. Everest. Venus has a wider variety of volcanic features per square mile than any other planet in the solar system.

There are also active volcanoes on other moons in the solar system. Io, a moon of Jupiter, may contain more active volcanoes than all the land areas on Earth. The key to Io's activity is called tidal friction, i.e. the pulling and pushing of Io's interior by the gravitational pull of Jupiter and its other moons. The surface of Io rises and fall some 150 feet each day! This forced movement creates frictional heat that comes out in eruptions.

The most common feature of a volcano is a conical shaped mountain, emitting lava and poisonous gases from a crater at its summit. However some volcanoes have different appearances. Shield volcanoes (so named for their shield-like profiles) are formed by the eruption of low-viscosity lava that can flow a great distance, but rarely explodes. They are common in the Hawaiian Islands and in Iceland. Lava domes are built by the slow eruptions of highly viscous lavas that do not flow far. Sometimes they form inside the crater of a previous volcanic eruption. Mt. St. Helens is such an example.

Then there are submarine volcanoes, which occur at the ocean's floor. If they occur in relatively shallow water their eruptions may break the surface of the ocean. Subglacial volcanoes occur beneath icecaps. When the ice melts, the lavas on the top usually collapse, leaving a flat-topped mountain. Examples of this can be seen in Iceland.

Volcanoes are also characterized by eruption types, from quiet lava emissions to explosive events. The kind of eruption is mostly affected by the magma composition and the amount of water present. It is beyond the scope of an introductory chapter such as this to detail these eruption types.

EFFECTS OF VOLCANIC ERUPTIONS

Volcanoes can kill people and destroy property through several different mechanisms.

First, in an eruption, volcanoes emit tephra (the name given to volcanic material ejected during an eruption). These emissions of rocks, magma and dust may be thrown into the air and travel many miles before coming to Earth, killing or injuring people on impact, and starting fires.

Second, the volcano can emit magma which moves along the ground destroying everything in its path. Third, the eruption can trigger lahars, landslides of water saturated debris coming down the side of a volcano. Fourth, if the eruption is explosive, their will be a pyroclastic flow of hot gases, dust, and particles that will burn to death, asphyxiate and/or bury anyone in its path. Fifthly, if the volcano is in or near water, its eruption may create giant tsunamis that will inundate coastal areas. In the last couple of centuries more persons have died from the tsunamis created by volcanoes than by all the other mechanisms combined.

Finally, if the eruption is truly immense, it hurls water vapor, carbon dioxide, sulphur dioxide, hydrogen chloride, hydrogen fluoride and ash into the stratosphere to heights of 10-20 miles. The sulfur dioxide is then converted into sulfuric acid which forms fine sulfate particles that reflect sunlight back into outer space thereby reducing the amount of sunlight reaching the Earth and thus cooling the Earth. The sulfuric acid may remain there for years, eventually falling back to earth as acid rain. These effects can be worldwide in scope.

SUPERVOLCANO

A supervolcano is a volcanic explosion with exponentially larger magnitudes than any volcanic eruption in historic times. It is sufficiently large to cover huge areas with lava and ash and cause long- lasting changes to the weather which in turn might threaten the extinction of the human race.

The supervolcano that most concerns U.S. scientists is the Yellowstone caldera. It is located in Yellowstone National Park in Wyoming, measuring about 34 miles by 45 miles. After a BBC program coined the term "supervolcano" in 2000, it has often been applied to the Yellowstone caldera. It is located on top of one of the planet's "hotspots." Within the past two million years it has undergone three extremely large explosive eruptions. Approximately 2.1 million years ago it emitted 2.5×10^{12} cubic meters of tephra, comparable to that emitted by Lake Toba, discussed below.

Chapter 5: Volcanoes

The next eruption occurred about 1.3 million years ago and the most recent about 640,000 years ago, which emitted 1×10^{12} cubic meters of tephra. In 2005 a television docudrama entitled Supervolcano, was televised. It depicted what would happen if Yellowstone erupted. According to the program, the eruption would cover virtually all of the United States with at least 1 centimeter of volcanic ash, cause mass destruction in the nearby vicinity and kill plants and wildlife across the country. In 2008 the Sci-Fi Channel aired the documentary, "Countdown to Doomsday" which included a segment called "Supervolcano."

Research shows that earthquakes that occur far away do affect the activity at Yellowstone such as the performance of geysers and hot springs. The federal government has classified Yellowstone as a "high-threat" system.

VOLCANIC EXPLOSIVITY INDEX

The Volcanic Explosivity Index (VEI) was devised in 1982 to provide a relative measure of the explosiveness of volcanic explosions. The volume of emissions, the height of the eruption cloud and other qualitative observations are used to determine the explosivity value. The largest volcanoes in history are given an "8." Such an eruption would have ejected at least 10^{12} cubic meters of tephra and have a cloud column over 25 kilometers high.

A value of "0" means that it was a non-explosive eruption with less than 10^4 cubic meters of material ejected. Each interval in the scale represents a ten-fold increase in these eruption criteria, except that going from a "1" (which represents at least 10^4 cubic meters of material ejected) to a "2" (which involves at least 10^6 cubic meters of material ejected) involves an increase of 100.

Chapter 5: Volcanoes

PAST VOLCANIC ERUPTIONS

We will here discuss several past eruptions. Obviously any such limited subset can only give the reader a brief glimpse into the destructive history of volcanoes.

Mount Toba in what is now northern Sumatra erupted about 74,000 years ago. It had a VEI of 8 and is classified as a supervolcano. It left behind a huge crater 50 miles long and 15 miles wide. The eruption left layers of dust 18 inches thick on ocean floors 1,500 miles away. The material ejected high into the atmosphere must have lowered the average temperature of the planet by perhaps 10 degrees Fahrenheit, making a climate that was already changing into another ice age even harsher for our human ancestors. One expert speculated that the eruption therefore contributed to the death of 60 to 75% of the human population worldwide.

Thera (today the island of Santorini, about 60 miles from the island of Crete in the Mediterranean Sea) exploded sometime in the period 1550 BC to 1630 BC. Crete at that time was the center of the Minoan civilization, a seagoing people who had built large palaces and engaged in overseas trade. Their culture had a high degree of organization. They were apparently involved in the making of bronze tools, ceramics, and luxury goods of gold and silver. Their cities were connected with stone-paved roads. Minoan buildings stood two and three stories high. Some even had plumbing facilities.

This advanced civilization (for its time) experienced a catastrophe from which it never recovered when the volcano on Thera exploded. The explosion ejected about 6×10^{10} cubic meters of material, giving it a rating of "6" on the VEI. It is believed that the explosion was several times more powerful than that of Krakatoa, to be discussed below. This should have deposited ash over part of Crete, but scientists have recently determined that no more than 0.2 inches

Chapter 5: Volcanoes

fell on most parts of the island so this was not the main cause of destruction of the Minoan society. Rather, a giant tsunami, that may have been higher than 100 feet, traveling at speeds of hundreds of miles per hour when it struck the Minoan coastal settlements, caused widespread death and destruction. The Minoans never recovered from that disaster and were soon conquered by the Greek state of Mycenae.

This catastrophe may be the basis of the Atlantis legend that an advanced civilization was inundated by a tsunami. It is also known that the sea level has risen by about 270 feet in the last 20,000 years as the glaciers on the continents melted. About 3,000 BC the sea level was 18 to 21 feet below the present level. Whether the Atlantis legend is based on this gradually rising sea level rather than a sudden catastrophic tsunami is a matter of conjecture.

Mount Vesuvius is an active volcano east of Naples, Italy. It is the only volcano on the European mainland to have erupted within the last hundred years. It has erupted many times in the historical era. Its most famous eruption was in 79 AD and led to the destruction of the Roman cities of Herculaneum and Pompeii. It has erupted about 42 times since then although none of these eruptions were as destructive as the 79 AD eruption. Its last eruption was in 1944. The eruptions are characterized by a plume dubbed a Plinian explosion, named after Pliny the Younger, the Roman writer who observed the 79 AD eruption. His uncle, Pliny the elder, died in it. The eruption column is estimated to have been 20 miles tall.

In early August of 79 AD springs and wells dried up. Small earthquakes started taking place on August 20, 79 AD, becoming more frequent over the next four days. On August 24 a catastrophic eruption started. The Plinian eruption lasted about 19 hours during which the volcano released about 4 cubic kilometers of rock and ash which engulfed both Pompeii and Herculaneum. It was then followed by two pyroclastic flows that engulfed both cities, burning and asphyxiating any stragglers who had not fled the city or been killed by the rain of

Chapter 5: Volcanoes

debris. Estimates of the population of Pompeii at that time range from 10,000 to 25,000 while that of Herculaneum was about 5,000. About 1,500 remains have been unearthed in the two cities, but the death total in the region must have been much larger.

The Mount Vesuvius volcano poses a threat in the future because of the density of the human population now living near to it. Plans have been made by Italian authorities for the evacuation of 600,000 people in the event that an eruption is anticipated. The plan assumes having at least a two week advance warning of an eruption. If authorities did not start an evacuation sufficiently in advance of an actual eruption, the death toll could be huge.

Mount Tambora is an active volcano on Sumbawa Island, Indonesia. Tambora erupted in 1815 with a rating of "7" on the VEI. It had roughly four times the energy of the Krakatoa explosion. The explosion was heard 1,200 miles away. It is the largest observed eruption in recorded history. The death toll was at least 71,000 people. Approximately 12,000 were killed directly by the explosion, probably caused by pyroclastic flows. Most of the other deaths were due to post explosion famine and disease. Only a moderate sized tsunami struck various islands in the Indonesian archipelago, with the largest reported being about 13 feet high.

The immense amount of debris, including sulphur, flung high into the atmosphere led to a worldwide cooling. Average global temperatures decreased by about 0.7 to 1.3 degrees Fahrenheit. This may not seem like much but the results were catastrophic. In the summer of 1816 countries in the Northern hemisphere experienced what was called "the year without a summer". In June 1816 snow fell in Albany, New York; such conditions continued for at least 3 months and ruined crops in North America. It is interesting to note that 1816 was the second coldest year in North America since 1400 AD. (The first coldest year, 1601, followed a volcanic eruption in Peru.) Europe suffered the worst famine of the 19th century.

Chapter 5: Volcanoes

Krakatoa is a volcanic island between Java and Sumatra in Indonesia. It exploded on August 27, 1883. It had a VPI of 6 and was equivalent to the detonation of 200 megatons of TNT--about 13,000 times more powerful than the atomic bomb which destroyed Hiroshima, Japan, at the end of WWII. The eruption ejected more than 2.5×10^{10} cubic meters of rock, ash and pumice. It also generated the loudest sound historically reported. The explosion was heard as far away as the island of Rodriguez, about 5,000 miles away. It destroyed 2/3 of the island of Krakatoa.

The Krakatoa volcano began erupting around July 20, 1883. On August 1 larger eruptions started. On August 24, eruptions further intensified. On August 26 the volcano emitted a black cloud of ash 17 miles high. At that point the eruption was virtually continuous. Ships up to 11 miles away reported pieces of hot pumice up to 10 centimeters in diameter falling on their decks. A small tsunami hit the shores of Java and Sumatra some 28 miles away.

On August 27, four enormous explosions took place with the last one being the loudest. Each was accompanied by a large tsunami, believed to have been over 100 feet high in places. Ash was propelled nearly 50 miles high. Around 1,000 people were killed by Krakatoa itself in and around the town of Ketimbang in Sumatra. Some researchers believe this was due to a lateral blast or pyroclastic flow, which crossed the water. The total official death toll given by Dutch authorities was 36,417, mostly from the giant tsunamis. Some sources think that the true death toll was 120,000.

The eruption had global consequences because of the cooling effect of so much material ejected into the upper atmosphere which reflected sunlight away from the Earth. Average global temperatures fell by as much as 2.2 degrees Fahrenheit. Temperatures did not return to normal until 1888.

The eruption darkened the sky for days afterwards, and

Chapter 5: Volcanoes

produced spectacular sunsets around the world for months following the eruption. Present day researchers suggest four possible causes of the explosions. First: the volcanoes' vents had sunk below sea level on the morning of August 27, 1883, letting seawater flood into it causing a massive series of explosions due to the interaction of ground water and magma. Second: the seawater could have chilled the magma causing it to crust over producing a pressure cooker like effect that was relieved only by explosions. Both of these theories assume that the island subsided before the explosion. However the pumice and other deposits are not consistent with a magma-seawater interaction.

The third theory is that a massive underwater land slump or partial subsidence suddenly left the highly pressurized magma chamber wide open. The final theory is that of magma mixing with a sudden infusion of hot basaltic magma into the cooler and lighter magma in the chamber below the volcano. This would have caused a rapid increase in pressure leading to the explosions.

In June 1927, a new island, named Anak Krakatau ("Child of Krakatoa") broke through the ocean water. Initially the eruptions of pumice and ash were washed away by the sea, but by August 1930, the new island broke water permanently and has since been growing in size. Since the 1950's the island has grown an average of 5 inches per week in height. The volcano's most recent eruptions were in October and November 2007. The book, <u>Krakatoa</u>, by Simon Winchester, gives an excellent description of the 1883 eruption.

Mt. St. Helens in the state of Washington erupted on May 18, 1980. Months before the eruption, the U.S. Geological Survey had established a base of operation at Vancouver, Washington to closely monitor the volcano. Survey volcanologist David Johnston was camping on Coldwater Ridge only a few miles from Mt. St. Helens. At 8:32 am Johnston radioed the USGS base and said "Vancouver, Vancouver, this is it!" The ensuing volcanic blast devastated the northern flank of the volcano, killing Johnston and 56 other persons.

Chapter 5: Volcanoes

Two other geologists were flying only 400 meters above the summit of Mt. St. Helens when the one of the largest landslides ever recorded in historical times took place followed by a massive explosion that shot out of the North side of the volcano towards Coldwater Ridge. They narrowly escaped death.

The blast flattened millions of trees in a 600 square kilometer area. This blast zone was also subjected to a huge debris avalanche, followed by the deposition of numerous lahars and pyroclastic flows. The debris avalanche partly filled Spirit Lake and then entered the North Fork of the Toutle River and flowed 23 km westward! The length of the avalanche makes it one of the world's largest ever recorded. Since the debris avalanche incorporated a large amount of water, voluminous lahars occurred later in the day.

Mt. St. Helens is about 37,000 years old, but has been especially active over the last 4,000 years, erupting at a rate of once per 100 years. The last eruption before 1980 had been 130 years earlier so some volcanologists had suggested that it was overdue. Historically Mt. St. Helens had typically generated explosive pyroclastic flows when it erupted, in contrast to other volcanoes currently active along the cascade chain, which usually generate non-explosive flows of lava.

Chapter 5: Volcanoes

FILM SUMMARIES

DANTE'S PEAK

Universal Pictures (US), color, 1997, 109 minutes

Cast: Pierce Brosnan (Dr. Harry Dalton), Linda Hamilton (Rachel Wando), Charles Hallahan (Dr. Paul Dreyfus), Elizabeth Hoffman (Grandma Ruth), Jamie Renee Smith (Lauren Wando), and Jeremy Foley (Graham Wando)

Director: Roger Donaldson

The film opens with the eruption of a volcano somewhere in Columbia where volcanologist Dr. Harry Dalton is fleeing with his wife Marianne. As they are leaving town a large piece of volcanic rock strikes their vehicle killing Marianne.

The film then continues four years later when Dalton is called back from his vacation to investigate volcanic activity in the town of Dante's Peak, located in the Northern Cascades in the state of Idaho. Meanwhile a pair of backpackers decides to take a dip in a hot spring there. Suddenly the hot spring is turned sulfuric from a small volcanic activity, killing the couple.

At the same time the town is receiving an award for being the second most desirable place to live in the United States, population under 20,000. It is also celebrating its annual "Pioneer Days Festival." There Dalton meets the town's mayor, Rachel Wando, who accompanies him with her two children, Graham and Lauren, to check the acidity in a lake near the peak. After checking the acidity and picking up the mayor's ex-mother in law, Ruth, the five of them head down to the town's hot springs where they discover the boiled bodies of the two backpackers. Just before discovering the bodies Harry

Chapter 5: Volcanoes

catches Graham as he attempts to jump into the hot spring.

Dalton then asks Wando to call a town meeting to discuss the option of putting the town on alert. As the meeting starts, Dalton's boss, Paul Dreyfus, arrives to evaluate the situation. He opposes putting the town on alert referring to an economically devastating decision made in 1980 when he had put another town on alert, but the nearby volcano did not erupt. Dalton, despite being urged to leave, decides to remain in order to help the US Geological Survey team to evaluate the possible eruption of the nearby mountain. Initially the volcano shows no sign of erupting and after a week the crew prepares to leave. Meanwhile Dalton forms a bond with the mayor and her children which turns romantic.

The night before the USGS team is to leave, the town's water supply turns acidic, due to the sulphur from the volcano breaching the town's springs. Dalton and Dreyfus then decide to put the town on alert and call for an evacuation. Wando attempts to get her ex-mother in law to come to town, but the woman stubbornly refuses to leave the bed and breakfast that she and her husband had built high up in the mountain.

During the town meeting alerting the townspeople to evacuate, the volcano starts to erupt leading to panic. Earthquakes collapse the town's main highway and traffic jams clog the only other route out of town. Meanwhile the children steal the Mayor's car and drive up the mountain to rescue their grandmother. When Dalton and Wando realize this, they head up the mountain to get the children just as the USGS crew prepares to depart. When they locate the grandmother she still refuses to leave. Her dog has run from the house into the woods and she and the children are searching for it. Down in the town, falling ash and rocks kill many. The rest of the residents are aided by the National Guard to escape.

Back up the mountain as Dalton and Wando argue with Ruth,

113

Chapter 5: Volcanoes

lava suddenly engulfs the house and all the vehicles, causing the five of them to flee to the nearby lake which they try to cross on a motorized boat. As they cross the lake they realize that the sulfuric acid in the lake is eating through the boat and its propeller. When they are almost across the lake the propeller fails completely, its blades eaten by the acid. Before it can reach the shore the boat begins to sink, so Ruth jumps off and pulls it to shore, but she is fatally burned by the acidity. The five of them then continue down the mountain on foot with Dalton carrying Ruth.

Back in town, the USGS crew is evacuated by the military, but a lahar (volcanic caused mudslide) causes the small bridge they have to cross to break apart. Two military vehicles barely make it across with most of the team but the third vehicle, driven by Dreyfus, is washed away, killing him. Up the mountain Ruth asks to be put down, and shortly thereafter dies. The other four continue down the mountain to a ranger station where they hot wire a truck and drive it towards the town. When they encounter a lava flow they find that it has mostly cooled on top and so they are able to drive on top of it. They also encounter the grandmother's dog who has also managed to survive by circling the lake.

When they reach the town they discover the washed out bridge. Realizing that they have no means of leaving the town, Dalton heads back to the deserted USGS headquarters and picks up a NASA tracking device which he places in the truck. He heads towards the town's mines and reaches a mine entrance just as a pyroclastic cloud is about to engulf the vehicle. Inside the mine Graham leads the way to the "hideout" he and his friends had made there, provisioned with food and water. Dalton then remembers that he had not turned on the NASA device, so he goes back to the truck. As he reaches the truck, there is a cave in and he is both injured and trapped inside the crushed truck. Eventually he reaches the NASA device and activates it. The device flashes for one or two days before its signal is noticed by the USGS team who think that Dalton had been killed by the pyroclastic cloud.

Chapter 5: Volcanoes

After being rescued from the mine by a massive rescue effort, Dalton learns that Dreyfus didn't make it but that "he got to see the show." On board a helicopter, Graham asks Dalton if he meant what he said in the mine about taking the family fishing. Dalton responds "sure did," clasps hand with Wando and kisses her as the helicopter flies over the ruins of Dante's Peak.

MOVIE TRIVIA

The director, Roger Donaldson, had been a geology major. He was thus very interested in the subject of the film.

The footage of the lahar coming over the dam was later used in *Killer Flood: The Day the Dam Broke* (2003).

VOLCANO

Twentieth Century Fox (US), color, 1997, 103 minutes

Cast: Tommy Lee Jones (Mike Roark), Anne Heche (Dr. Amy Barnes), Gaby Hoffman (Kelly Roark), Don Cheadle (Emmit Reese), Jacqueline Kim (Dr. Jaye Calder)

Director: Mick Jackson

Mike Roark is a divorced Los Angeles emergency official who is on vacation with his daughter Kelly as the film opens. There is an earthquake registering about 5 on the Richter scale so he returns to work. Seven workers in an underground tunnel are killed by an underground steam eruption of unknown origin. When Roark and another worker go back down into the same tunnel they find it very hot in the tunnel and discover that the heat is due to magma which is moving under the ground. This occurs near the La Brea Tar Pits which

Chapter 5: Volcanoes

begin to warm rapidly.

Roark seeks help from scientists to explain what is going on. Dr. Amy Barnes then enters the story. She is a geologist who wants to return to the tunnel to take readings. Roark forbids this as being too dangerous. As a result Dr. Barnes and an assistant return to the tunnel at 4 AM the next morning when the police are gone. While they are in the tunnel a more powerful earthquake occurs opening a crack in the floor of the tunnel and resulting in the death of Dr. Barnes' assistant. The earthquake is centered 60 miles away from Los Angeles but it nonetheless opens fissures in the Earth which brings magma to the surface.

A volcano erupts in Wilshire Boulevard and hurls huge chunks of debris into the air, killing people and destroying property. Meanwhile Roark's daughter, who has been burned in the eruption, goes with a Doctor Jaye Calder to Cedars Sinai Hospital while her father organizes a response to the moving lava.

The lava flow goes down Wilshire Boulevard, through the Metro Red Line subway tunnel, and forms a fountain of lava near to the Beverly Center Shopping Mall. The lava destroys one subway train stranded in the subway by the earthquake, burns cars, firemen, politicians, and damages much property. Roark tries to stop the flow of lava by building a concrete barrier and dropping large amounts of water on it as it reaches the barrier so as to solidify the surface. This succeeds and Roark and his colleagues celebrate.

Roark then learns from Barnes that the lava is moving in another direction under San Vincente Boulevard. He tries channeling the lava flow into the ocean via the concrete channel of Ballona Creek. However, he is reminded by Dr. Barnes that San Vincente Boulevard slopes in the wrong direction and will send the lava into Cedars Sinai Hospital where his daughter is being treated. Thus charges are planted to change the channel diverting the lava.

Chapter 5: Volcanoes

When lava erupts from a block in San Vincente Boulevard, the new 22 story Beverly Hills apartment building right across the street is brought down safely diverting the lava to the sea and saving Los Angeles. Meanwhile Roark saves his daughter and a little boy named Tommy from death as the apartment building is brought down. Roark and his daughter then leave for a second attempt at a vacation, driving away with Dr. Barnes.

MOVIE TRIVIA

Volcano was released in the United States on April 25, 1997, just months after *Dante's Peak* which opened in February 1997. Its box office gross was $120.1 million worldwide, considerably less than the $178 million world wide gross of Dante's Peak.

Volcano received a rating of 35% on RottenTomatoes.com, a rather poor rating. Commentators comparing the two films favored *Dante's Peak*.

SCI-FI SCIENCE vs. REAL SCIENCE

LOCATION OF THE VOLCANO

DANTE'S PEAK. The movie is set in the Cascade Range in the state of Idaho. However there are no Cascades or other active or dormant volcanoes in Idaho. However there are such volcanoes in the nearby states of Washington and Oregon.

VOLCANO. The closest tectonic plate to Los Angeles is the San Andreas Fault which is where two tectonic plates slide past one another. This does not produce volcanoes. There are also no known geologic hot spots for hundreds of miles from Los Angeles.

Chapter 5: Volcanoes

TYPE OF ERUPTION

DANTE'S PEAK. Volcanoes in the Cascade Range frequently do erupt explosively and do produce pyroclastic flows as depicted in the film. They also produce lahars as shown in the film.

VOLCANO. The eruption is mainly a lava event with only a little material ejected skyward.

THE WARNING OF AN IMPENDING ERUPTION

DANTE'S PEAK. The relatively short warning of an eruption depicted in Dante's Peak has actually occurred. The eruption of a volcano in Alaska on December 13, 1989 was preceded by only 24 hours of earthquake activity. However some volcanoes have exhibited earthquake activity for months without erupting.

VOLCANO. Again, volcanoes have erupted with little warning, as was the case here.

LAVA FLOW

DANTE'S PEAK. Fast flowing lava usually only erupts from Hawaiian or "shield" volcanoes, not from a Cascades volcano which usually emits thick slow moving lava that rarely moves far from the vent. The volcano dome is a mound of rather thick such lava. Volcanoes like those in the Cascades rarely produce both pyroclastics and lava flow in the same eruption.

VOLCANO. We are shown both slow moving and fast moving lava coming from the same source. This seems unrealistic.

Chapter 5: Volcanoes

COULD A CAR, BUS OR TRAIN TRAVERSE A LAVA FLOW?

DANTE'S PEAK. Lava is very hot, often at a temperature over 1500 degrees Fahrenheit, and rubber, wood and people brought near the lava would burst into flame. The radiative heat from the lava would ignite flammable materials without even touching the lava. Probably the car would only last seconds on a lava flow before bursting into flames, killing its occupants.

VOLCANO. The sequence in which a rescuer carries a man slowly through the inside of the stranded subway train car, below which lava is flowing, and then holds him while jumping into the lava and then finally throws him to safety is unrealistic. Both would have died from the heat inside the train car.

ACID LAKE

DANTE'S PEAK. Lakes near volcanoes can become acidic and the acid could corrode metal parts as depicted. However the time scale seems much too short for the acid in the lake to disable the boat as depicted.

HOT SPRINGS BOILING SWIMMERS ALIVE

DANTE'S PEAK. The hot springs in a geothermal area (e.g. Yellowstone National Park) can change temperature abruptly due to earthquakes that change the underground systems of cracks in the rocks that control the flow of hot water. However hot springs do not heat up so quickly as to kill the couple in it in a matter of seconds.

TREES AND ANIMALS KILLED BY CARBON DIOXIDE RELEASED BY MAGMA

Chapter 5: Volcanoes

DANTE'S PEAK. Carbon dioxide released from underground magma chambers can accumulate in the soil and kill plants. It can also suffocate people and animals as the carbon dioxide is heaver than air which it pushed upward causing suffocation near ground level.

VOLCANO. The occupants of the stranded subway train are found unconscious. Whether they collapsed due to the heat or the carbon dioxide is unclear.

VOLCANIC TREMORS

DANTE'S PEAK. Earthquakes or volcanic tremors all feel the same to a person but will look different on the tracing made by a seismograph, which records the amount of ground shaking produced by an earthquake or volcano. Volcanic quakes usually register only 4 or 5 on the logarithmic Richter scale. The destruction in the movie caused by the volcanic quake appears to be much greater, perhaps a 6 or 7 magnitude earthquake.

VOLCANO. The tremors depicted in the film have struck California cities in the past (and caused even more damage than depicted). However those earthquakes were not caused by a volcanic eruption nor did they precede one.

LOSS OF LIFE

DANTE'S PEAK. It is not clear how many townspeople perish in the eruption, but clearly most are evacuated safely so that the death total should have been only in the tens of persons. This is consistent with the death of only 57 persons when Mt. St. Helens erupted.

VOLCANO. The film says that only a 100 people died with thousands injured. This seems incredibly low for a volcano erupting without warning in a city of millions.

Chapter 5: Volcanoes

QUESTIONS

1. Which of the two films is more accurate scientifically? Support your answer with specifics.

2. Why is it unlikely for a volcano to occur under Los Angeles?

3. In *Volcano* how close to the moving magma do people stand? Is this realistic?

4. Do dogs and/or cats behave oddly just before an earthquake?

5. Has Roark instructed his daughter properly on what to do during an earthquake?

6. Do volcanologists plant seismographs in or near a volcano's cone shaped top as depicted in *Dante's Peak*?

Chapter 5: Volcanoes

7. Would the NASA tracking device have been detected as depicted in *Dante's Peak*?

8. In reality would the passengers of the truck in *Dante's Peak* have survived the pyroclastic cloud by crashing through the entrance to the mine?

9. Was there anything that Dr. Dalton could have used to paddle to the shore of the acidic lake, thereby making it unnecessary for the grandmother to jump into the lake to pull the boat the last few feet?

10. What do you think of the actions of Dr. Barnes in going into the tunnel with her assistant even after being told that it was too dangerous?

CHAPTER SIX

PANDEMICS AND MODERN PLAGUES

THE SCIENCE OF *OUTBREAK* AND *THE ANDROMEDA STRAIN*

The biological threats to humanity are legion and likely have been with us for the duration of our existence. Evidence of their presence and effects in antiquity has been found in mummies from pre-historic times when no written account of disease was made. It appears in ancient literature, including The Bible. The pestilence God sent upon Egypt in the account found in the book of Exodus and depicted in the film, *The Ten Commandments*, includes death to the animals of the Egyptians and finally death to the first-born of both the Egyptians and their cattle. In a more recent scourge, the Black Death of the fourteenth century claimed the lives of an estimated 33 – 67% of the people in Europe. The early twentieth century witnessed an influenza pandemic that coincided with World War I. Earlier epidemics and pandemics remained confined to limited geographic areas. The frequent intercontinental movements of humans over long distances and to diverse locations during World War I helped to transmit the disease and insure its global distribution, thus creating a pandemic that was intercontinental.

More recently in both the twentieth and current centuries, influenza continues to occur pandemically. In addition, the human immunodeficiency virus, HIV, has spread throughout the world in pandemic proportions. Clearly biological threats were important historically and remain important on the contemporary health scene.

Chapter 6: Pandemics and Modern Plagues

Despite more tools than ever to deal with illness, biological threats arguably appear to be on the brink of turning the tables on humanity and are likely to be of increased significance in human affairs in the foreseeable future. These and other issues will be explored in the context of the films, *Outbreak* and *The Andromeda Strain*.

EPIDEMICS AND PANDEMICS

An <u>epidemic</u> is the occurrence of a disease among many members of a community in a continuous period of time. This pattern is in contrast to that of an <u>endemic</u> disease, which is found locally normally and produces illness regularly, but with smaller numbers of cases at any given time. Endemic disease produces small outbreaks, either continuously or episodically, as with seasonal illness. An epidemic can be due to the introduction of a disease from outside the immediate region or to an unusual, temporary, increase in the cases of an endemic disease. Introduction of a disease by travelers can produce serious epidemics when the disease introduced is completely new to the area, rather than simply a new strain of an agent already present. For example, the arrival of Europeans in North America brought illnesses to Native Americans that were not endemic, such as measles, syphilis, and tuberculosis. The death rate was high and illness was severe in those who survived. Because these diseases were not endemic, the Native Americans had not developed immunity to them.

A <u>pandemic</u> is the occurrence of a disease over a large geographic region. Often it affects more than one country or more than one continent, but that need not be the case. The World Health Organization has established phases for influenza pandemics to characterize the capacity of the virus to produce pandemic (See Table 1). An actual pandemic is Phase 6.

Chapter 6: Pandemics and Modern Plagues

PANDEMICS OF THE PAST

Plague. Plague swept through the human population of Europe over the period from 1347 to 1351. Believed to be introduced into humans by the bite of a flea, *Yersinia pestis*, the plague bacterium, is traditionally named as the pathogen of the Black Death. Manifesting in two major forms, bubonic and pneumonic, the disease killed large numbers of people. The deaths in Europe have been estimated at around 20 million, and the total for the pandemic has been estimated at 75 million. The pandemic originated in China and spread westward with trade. The Middle East and Northern Africa, as well as Europe, were devastated by the disease. While bubonic plague is probably the better known form of plague, pneumonic plague, which infects the lungs, is more problematic. In bubonic plague, the lymph nodes swell and produce the protrusions known as buboes. In this form, plague is not transmissible from person to person. Person-to-person transmission is required for a pandemic. When plague infects the lungs, it can be coughed into the air in tiny droplets and thus be transmitted to other people without any actual personal contact. The Black Death was first attributed to *Y. pestis* in the late nineteenth century. Recently, skepticism that the fourteenth century pandemic was caused by *Y. pestis* has emerged. Among the alternatives proposed are pulmonary anthrax and an Ebola-type virus.

Influenza. Influenza is caused by a virus. While it produces many symptoms that occur outside of the respiratory system, it is fundamentally a respiratory disease. Most of the symptoms are due to the inflammatory response that occurs as part of the immune response to the virus. An inflammatory response that succeeds in controlling a virus will be brought to a halt. One that fails to do so will persist and, if it grows stronger and stronger, ultimately become life-threatening itself. Historians have identified numerous pandemics throughout recorded history that match the symptoms of influenza.

Chapter 6: Pandemics and Modern Plagues

One of these occurred in 412 B.C. and was described by Hippocrates. In the influenza pandemic of 1918, sometimes called the Spanish Flu, 50 to 100 million people are estimated to have died worldwide, over a period of 18 months. This estimate has a large range which can be attributed to the diverse locales with varying practices in public health record-keeping and to the reluctance of some countries to be forthcoming about disease mortality during wartime. By recovering samples from bodies of victims of the 1918 pandemic, scientists have identified the influenza strain as H1N1 (See p. 129.). This outbreak of influenza was peculiar in both its time of occurrence and its human targets. Influenza typically kills young children, whose immune systems are still developing, the elderly, whose immune systems no longer function optimally, and those with illness that weakens the immune system. The remainder of the population is less likely to die from influenza. The Spanish Flu killed many apparently healthy people from the remainder group that is normally less likely to die from an influenza infection. The 1918 pandemic was at its height in the summer and fall. This timing is peculiar because influenza is usually most active in the winter months. H1N1 influenza has recurred on multiple occasions since 1918 (See Table 2.) without the odd characteristics it displayed in 1918.

Because Temple University, which will be the first to use this text, is located in Philadelphia, Pennsylvania, that city and state will be used to illustrate the impact of the 1918 influenza pandemic. Secretary of Health and Human Services, Mike Leavitt, made these remarks about the impact of mass illness on the city at the Pennsylvania State Summit on March 20, 2006:

On September 27, 1918, Pennsylvania optimistically reported that 'comparatively few cases' had been reported among the civilian population. Then influenza took hold. On October 4, the state reported that the disease was epidemic in Pittsburgh and Philadelphia. Nearly 15,000 cases were counted in the first 18 days of October, and the dreadful toll continued to climb. Philadelphia was one of the hardest hit cities in United States. As the disease spread, essential services collapsed. Nearly 500 policemen failed to report for duty. Firemen, garbage collectors, and city administrators fell ill. The city's only morgue overflowed. It was built to handle 36 bodies, but contained more than 500. Bodies accumulated in the morgue's hallways and lay there rotting. Five supplementary morgues were eventually opened. Convicts were recruited to dig graves. There were never enough coffins, and people would steal them from undertakers when they could. Public gatherings were banned to restrict the spread of the disease. Streetcars were shut down. Schools, churches, and places of public meeting were closed, and so were theaters and places of amusement. . . . Nearly 24,000 Pennsylvanians died during the first month of the disease. By October 25, after the first wave of the pandemic had passed, it was estimated that 350,000 people had been struck with the flu (about 150,000 of whom were Philadelphians).

POSSIBLE PANDEMIC CATASTROPHES OF THE FUTURE

VIRUSES

The contemporary world has had glimpses of pandemic devastation, but so far no pandemic has captured the attention of the public and the powerful, in politics and other walks of life, so as to demand solutions. The current pandemic of HIV AIDS and the prospect of "bird flu" have come closer than other illnesses in reaching that level of concern. Both are viral diseases.

An agent that causes disease is called a <u>pathogen</u>. Most pathogens are microbial. The most troublesome of the microbes for humans are viruses and bacteria. Other microbes that cause human disease are fungi and protozoa. The most common of the pathogens that are not microbes are worms. A <u>virus</u> is a very small molecular complex of nucleic acid, protein, and, sometimes, lipid and carbohydrate. Viruses cannot reproduce on their own and require the use of cells to do so. A virus uses molecules on its surface to gain entrance to a cell, referred to as the host cell. Viral surface molecules have spatial configurations that mimic molecules that would normally bind to molecules on the surface of the host cell. Once inside the host cell, the virus uses the biosynthetic apparatus of the cell to make new viral particles (<u>virions</u>). The instructions for viral reproduction are contained in the virus's genetic material, which can be either DNA (deoxyribonucleic acid), as it is in a human cell, or RNA (ribonucleic acid).

    HIV AIDS. <u>HIV</u> is the <u>human immunodeficiency virus</u>. It causes <u>AIDS</u>, acquired immunodeficiency syndrome. HIV AIDS is distributed worldwide with the highest case numbers occurring in Africa. The disease takes a variable period of years from the time of infection with the virus to the time of death from the immunodeficiency it causes. A rapid progression may cause death relatively quickly in a year or two, while a slow progression, especially with improved medication, may take ten to twenty years. This lag between infection and death, which produces greatly staggered occurrences of death relative to the time of infection, has blunted the impact of the disease, and, despite the large numbers of infected people and the immense cost of treating them, the AIDS pandemic has arguably not maintained the urgency that was generated when the disease was first recognized in the 1980s.

    The immunodeficiency caused by HIV arises from its ability to infect one type of lymphocyte called a <u>helper T cell</u>. Lymphocytes are one of the five types of white blood cells (neutrophils, lymphocytes, monocytes, eosinophils, and basophils). The human immune system depends heavily upon helper T cells for effective immune responses. With an HIV infection, the helper T cells die in large numbers. For a time the helper T cells can be replaced by the bone marrow.

Chapter 6: Pandemics and Modern Plagues

Meanwhile, the immune system tries to fight the virus, but frequent viral mutations ultimately defeat the immune response. When the levels of helper T cells fall far enough below normal, people with HIV AIDS have an immune system that underperforms, and they are said to be immunocompromised. They then become susceptible to infections that a healthy immune system would prevent. Such infections are called opportunistic infections. Some examples of opportunists are *Pneumocystis carinii*, *Mycobacterium avium intracellulare*, cytomegalovirus, and *Candida* species (yeasts).

Bird Flu/Influenza. Asia historically has been the birthplace of new strains of influenza, and bird flu is no exception. The current strain of avian influenza was first recognized in 1996. It was isolated from a goose in Guangdong, China, and identified as an H5N1 strain. Influenza strains are identified by the molecular type of two major molecules that are important in the recognition of the virus by the immune system. These are hemagglutinin, represented by capital H and a strain number, and neuraminidase, represented by capital N and a strain number. Because influenza infects a number of species besides humans, in settings where the different species are in proximity, there is an opportunity for a cell to be infected by a virus particle (virion) that is a strain from another species as well as a virion of a human strain. This can lead to exchange of genetic material as virions are assembled in the cell and, in the process, create new strains. Such a change is called an antigenic shift. When an antigenic shift occurs, the strain designation changes. For example, the Hong Kong Flu which appeared in 1968 was H3N2. Eleven years later the Bangkok Flu of 1979 was H3N2, also. In the preceding year, 1978, the H1N1 Brazil strain prevailed. Thus, both the H and N antigens underwent antigenic shift in 1978 and then rapidly shifted again between 1978 and 1979 to return to the same strain type as was prevalent in 1968.

The H5N1 avian influenza is presently producing very high mortality among bird populations that it infects. Table 2 shows that the

Chapter 6: Pandemics and Modern Plagues

H5 hemagglutinin has been recognized in avian influenza for many years. The tern strain isolated in 1961 was H5N3. Notice that no H5 strain has been isolated from a human epidemic. The H5 character is new to our species. Therefore, no humans had immunity to it when it first appeared. This suggests that if an H5 strain becomes sustainably transmissible from human to human it could be devastating. So far transmission has been from poultry to humans, and no sustained human-to-human transmission is known.

For reasons that are not understood well, some pathogens produce much stronger inflammatory responses than others. This becomes a problem if the immune response fails to limit the amount of pathogen present in the body. When that occurs, some of the chemicals that produce inflammation, which belong to the class of bioactive molecules called <u>cytokines,</u> are produced in unusually large amounts for continuous periods of time. The result is the excessive, life-threatening inflammation that is called a <u>cytokine storm.</u> The inflammatory response is complex and over one hundred cytokines are known to play a role in it. The precise mix of cytokines and the dominant ones depend upon the particular pathogen and the location of the infection. The H5N1 bird flu in humans is known to kill by producing a cytokine storm. Research comparing H5N1 and H1N1 in cell culture studies has demonstrated that H5N1 is a more powerful inducer of pro-inflammatory substances than H1N1. Pro-inflammatory substances promote inflammation, unlike anti-inflammatory substances, which reduce it. Among the pro-inflammatory cytokines released during H5N1 infections are TNF-α (tumor necrosis factor alpha), IFN-β (interferon-beta), IP-10 (interferon-gamma-inducible protein-10), IL-6 (interleukin 6), and RANTES (regulated on activation, normal T cell expressed and secreted). At one stage in the investigations of H5N1 influenza, researchers hoped that lives could be saved if the strength of the inflammatory response could be controlled. It now appears that the solution is not that straightforward.

Chapter 6: Pandemics and Modern Plagues

<u>Ebola Hemorrhagic Fever.</u> This is the disease on which *Outbreak* is based. It was presented as a possible agent of global pandemic in the book, <u>The Hot Zone</u>, by Richard Preston. He and others have explored the possibility that some new pathogens have been introduced to human populations when hitherto undisturbed habitats, such as deep, pristine jungles, have their previously isolated pathogens become accessible to humans when land is cleared for human use or humans live in proximity to jungles that previously had no humans near them. When pathogens move from species to species, they do not necessarily produce the same severity of illness. In some cases, a microbe can cause mild disease or no disease at all in one species, but cause serious illness in another. Such a situation is suspected often in the case of new pathogens. The species that harbors the pathogen is called the <u>reservoir</u>. In the case of the Ebola virus no reservoir has been identified definitively. Many think the most likely candidate is a bat, since fruit bats given Ebola develop an infection that is not life-threatening, as is typical of a reservoir species. The search to identify the reservoir continues, with mammals, especially rodents, and arthropods the foci of the investigations. The Ebola virus is transmitted by contact with virus-laden body fluids.

Initial symptoms of the illness are common to numerous illnesses and include fever, aching muscles, joint pain, headache, malaise, weakness, sore throat, nausea, and dizziness. The devastating internal events are not completely understood. Either by a virus-induced disruption of the balance in blood clotting regulation or due to a strong inflammatory response, clots form in small blood vessels. This is called <u>disseminated intravascular coagulation</u> or <u>DIC</u>. It renders highly vascular organs that receive a large blood supply in danger of organ failure when a large fraction of their tissue loses its blood supply because blood vessels are blocked. Thus, among the organs that often fail in an Ebola infection are the liver, spleen, and kidneys.

Chapter 6: Pandemics and Modern Plagues

BACTERIA

<u>Drug-resistant bacteria</u>. For decades, humans have had their lives extended by the use of purified and synthetic antibiotics. In as few as two generations, the fear of illnesses that had frequently killed has been lost. People with a diagnosis of bacterial pneumonia now expect to be prescribed antibiotics and to recover. Before the use of penicillin, bacterial pneumonia was often a killer. The bacteria, however, have resilience of their own. Many strains became resistant to penicillin.

The pharmaceutical industry answered with chemically modified penicillins as well as combination therapy that includes a chemical to block a bacterial penicillin-digesting enzyme that is found in some resistant strains. Numerous human bacterial pathogens have the ability to develop resistance to antibiotics. The means to deal with that has typically been to urge restraint in the use of antibiotics and to try to use some antibacterials sparingly so that resistance to them would be unlikely. That approach may be reaching the limit of its success. Now even the antibiotics that were reserved for difficult cases are failing because of resistance to them.

A prime example of new resistance is <u>vancomycin resistance</u>. Vancomycin used to be the drug that would succeed when resistance was encountered. Physicians reserved it for use on difficult cases involving bacteria classified as Gram positive. Vancomycin has to be administered intravenously because it is not absorbed from the gut. It is a powerful chemical. Its side effects include damage to the kidneys and the ears. People who are treated with vancomycin find those risks preferable to dying from their bacterial infections. Unfortunately, more and more, there are resistant strains that overcome vancomycin.

Another type of antibiotic resistance is found in strains of *Staphylococcus aureus*, a bacterium that is often involved in infections

Chapter 6: Pandemics and Modern Plagues

obtained in a hospital. These strains were designated as methicillin-resistant *Staphylococcus aureus* or MRSA. Since resistance to other antibiotics occurs as well, MRSA is also used to mean multiple-resistant *Staphylococcus aureus*.

These problematic strains are resistant to antibiotics that are classified as β-lactams, which includes the penicillins and the cephalosporins. So far these bacteria have been dealt with both by traditional means, such as susceptibility testing to find a useful antibiotic, and unconventional means, such as the use of maggots and bacteriophage therapy.

Now the race is on. For the better part of a century the health care community has been winning for the humans. But now, the bacteria are gaining. The winner of this heat is not yet clear.

PRIONS AND VIROIDS

Prions are infectious particles that are made only of protein. Prions cause neurodegenerative diseases. The first to be discovered was scrapie, a disease of sheep. Prion disease is known in several other mammalian species. The most notorious is bovine spongiform encephalopathy or mad cow disease. In humans the diseases Creutzfeldt-Jakob disease, fatal familial insomnia, Gerstmann-Straussler syndrome, and kuru are attributed to prions. Examples of medical transmission of prions are via contaminated surgical instruments, corneal transplants, and cadaveric growth hormone.

Viroids are infectious particles that are made only of ribonucleic acid, RNA. They are common plant pathogens, such as the potato spindle tuber and the coconut cadang cadang viroid. The only known human illness caused by a viroid is hepatitis D. It requires a co-infection with the hepatitis B virus. The viroid becomes transmissible when it uses a hepatitis B coat called the capsid. In that

133

Chapter 6: Pandemics and Modern Plagues

form, like hepatitis B, it can be transmitted by blood. The enzymatic activity of the hepatitis D viroid leads to the death of liver cells. Because its multiplication requires another virus, some prefer to call hepatitis D a subviral particle or a subviral satellite.

Prion illness challenges established notions of infectiousness. Central to the concept of creating disease by the introduction of microbes in an individual is the ability of the microbe to reproduce itself using a nucleic acid genome of DNA or RNA. Prions have no nucleic acid. Moreover, human prion diseases are known to occur both sporadically, apparently by transmission, and genetically in families. The prion protein is encoded in the genome of the afflicted individual. This is unlike any traditional infectious disease, in which the infectious agent has its own distinctive genome and is never a normal part of the host genome.

The claim that prions cause illness has been challenged repeatedly. An interesting counterclaim suggests that the real agent of human disease blamed on prions is the spiroplasma. Spiroplasmas, like mycoplasmas, have no cell wall. They can be thought of as minimal bacteria because they have one of the smallest known genomes. Dr. Frank Bastian, a proponent of this idea, finds persuasive that experimental spiroplasma brain infections cause symptoms similar to scrapie and that the internal spiroplasma fibrils are very similar to scrapie-associated fibrils (SAF), to the point of cross-reacting with antibodies to SAF. Cross-reactivity means that antibodies, which make very specific recognition of molecular configurations, recognize not only the material that was used to elicit their formation, but also something else. In this case the antibodies formed in response to SAF are able to bind to spiroplasma fibrils, which indicates important similarity in their configurations. Research that led to the discovery of prions, however, found that no nucleic acid was necessary for the transmission of brain disease. If spiroplasmas were transmitting disease, nucleic acid would be expected since the spiroplasmas contain both DNA and RNA.

STEALTH INFECTIONS

Infectious diseases that take years to develop and are not usually diagnosed until they are quite advanced have been called slow infections. The term has been applied to diseases caused by prions. More recently the term stealth infection has come into use by some. It suggests that slowly developing infections without early overt symptoms normally associated with infectious disease are sneaking up, not just on individuals, but also on society.

The association of cardiovascular disease, not just with one's genetic heritage or one's diet, but in some cases with bacterial infections by organisms such as *Chlamydia pneumoniae*, is an example of a stealth infection. Hanna Kälvegren and others have argued that *C. pneumoniae* infections lead to the formation of reactive oxygen species, and that they, in turn, cause oxidation that promotes arteriosclerosis. The claim has been made that such illness is a greater threat to human societies than the traditional agents of epidemics and pandemics.

BIOTERRRIORISM AND BIOLOGICAL WARFARE

The use of biological agents against enemies during a war has been feasible on a large scale for decades. Fortunately, since biological agents indiscriminately target both combatants and non-combatants, it has not occurred on a large scale. The treaty, "Convention on the Prohibition of the Development, Production and Stockpiling of Bacteriological (Biological) and Toxin Weapons and on Their Destruction," the Biological Weapons Convention for short, signed by 156 parties, including the United States, now bans many types of biological warfare. Signatories agree that they will not "develop, produce, stockpile or otherwise acquire or retain: microbial or other biological agents, or toxins whatever their origin or method of production, of types and in quantities that have no justification for prophylactic, protective or other peaceful purposes; Weapons, equipment or means of delivery designed to use such agents or toxins for hostile purposes or in armed conflict." The Biological Weapons Convention has been in force since 1975.

Bioterrorism, for the most part, operates outside of the reach of diplomacy. There is no reason to expect that it is excluded from the range of possible attacks by terrorists. The attempt to kill members of Congress with anthrax immediately after the September 11, 2001, terror attacks on the twin towers of the World Trade Center of New York City raised the awareness of Americans to this possibility.

The Centers for Disease Control (CDC) recognizes three categories of bioterrorism agents, Categories A, B, and C. (See Table 3.) Category A includes agents, including toxins, that cause serious illness with high mortality and that can be disseminated on a large scale. Their use would be expected to produce widespread infection that would cause panic and social disruption. Category B agents have low mortality and are moderately easy to spread. Category C agents are emerging pathogens with the potential to be developed into bioterror agents with high morbidity and mortality. They are easy to produce and spread.

Chapter 6: Pandemics and Modern Plagues

FILM SUMMARIES

OUTBREAK

Warner Brothers (US) 1995, color, 127 minutes

Cast: Dustin Hoffman (Sam Daniels), Rene Russo (Robby Keough) Morgan Freeman (Gen. Billy Ford), Kevin Spacey (Casey Schuler), Cuba Gooding, Jr. (Maj. Salt)

Director: Wolfgang Peterson

An outbreak of hemorrhagic fever occurs in the jungles of the Motaba Valley in Zaire (now the Democratic Republic of the Congo) in 1967. Medical personnel land and investigate it. A whole village has rapidly become infected with a disease that kills a person in 2 to 3 days. The military decides to wipe out the village to prevent further infection. The village is destroyed by a fuel-air bomb, the most powerful non-nuclear weapon in the U.S. arsenal.

Some 30 years later the virus surfaces again in Africa. Colonel Sam Daniels, an army M.D., and his team are sent to Zaire to investigate. They succeed in containing the virus and return to the U.S. Daniels asks his superior, General Ford, to put out an alert on the virus. Ford, who had gone to the Motaba Valley in 1967, recognizes the outbreak as a reoccurrence of the Motaba virus. He tells Daniels that it is unlikely to show up in the U.S. and refuses to issue an alert.

The military doctors and the CDC are unaware that a host of the virus was captured in the Motaba Valley and imported illegally. The exotic animal pet trader, Jimbo Scott, takes the monkey to a pet store in Cedar Creek, California. The monkey infects both Jimbo and the pet store owner. Since the pet store owner refuses to keep the monkey because he wanted a male, Jimbo releases the monkey into the

Chapter 6: Pandemics and Modern Plagues

nearby woods. Then he develops illness during a flight to Boston to meet his girlfriend. He infects his girlfriend, they are both hospitalized and placed in isolation, but soon they die. The CDC is notified and sends Daniels' ex-wife, Dr. Robby Keough, to investigate. The Boston outbreak is successfully contained.

Meanwhile, in Cedar Creek, the infection is being spread. The CDC is on hand to oversee and contribute to the efforts to contain the infection. Robby heads the CDC team. Daniels learns of the outbreak and wants to participate in the investigation. Fearing his dogged persistence, his superiors want him out of the way and order him to New Mexico for another assignment. He disobeys and flies to Cedar Creek.

The investigators recognize the significance of the afflicted pet store owner and suspect that the man was infected by a host animal. They surmise that the animal may have antibodies to an airborne strain of Motaba. The size and rapid spread of the outbreak lead to a military quarantine of Cedar Creek. The citizens who try to flee learn that it is a serious matter when some are shot after refusing to heed a command to turn back.

Accidents involving breaks in safety techniques lead to the infection of Robby and Casey. Sam Daniels is more determined than ever to find a cure with the life of his ex-wife, whom he still loves, at stake. When General Ford provides antiserum E-1101 to try on people infected with Motaba, Daniels decides to try it on an infected monkey from the pet store as well. The monkey gets better; the people do not. Daniels concludes that the antiserum was not a newly produced experimental product, but rather was pre-existing. He confronts General Ford, who admits that the antiserum traced to the original 1967 outbreak, which he says was kept secret for reasons of national security.

Chapter 6: Pandemics and Modern Plagues

Soon Daniels learns of Operation Clean Sweep, a military plan spearheaded by General McClintock, to destroy all of the inhabitants of the town of Cedar Creek with a fuel-air bomb, as was done in Zaire. Daniels begins to search frantically for the original host animal. He succeeds and takes the monkey to the mobile lab at Cedar Creek. Major Salt is then able to prepare antiserum. Daniels next learns that Operation Clean Sweep is underway, but he succeeds in persuading pilots of the bomber to reconsider their actions. They release the bomb over the water rather than the town. General Ford has had his fill of the apparently unnecessary secrecy that threatened the lives of so many. He places General McClintock under arrest for withholding information from the President, who had approved the bombing of Cedar Creek. The citizens of Cedar Creek receive antiserum to stop the outbreak, and Sam reconciles with his ex-wife, Robby.

THE ANDROMEDA STRAIN

Universal Pictures (US), 1971, color, 131 minutes

Cast: Arthur Hill (Dr. Jeremy Stone), David Wayne (Dr. Charles Sutton). James Olson (Dr. Mark Hall), Kate Reid (Dr. Ruth Leavitt)

Director: Robert Wise

The film tells the story of four days of crisis arising from the failure of a satellite, launched in Project Scoop, whose purpose was to collect pathogens that might be useful in biological warfare; it fell out of orbit and crashed near the small town of Piedmont, New Mexico. When airmen are sent to Piedmont to recover the satellite, they discover that almost everyone is dead.

Chapter 6: Pandemics and Modern Plagues

Figure 5. Outbreak: Scientists in protective gear entering a village which is infected with a lethal virus. Courtesy of Warner Brothers/Photofest.

Subsequently a Wildfire alert is issued, and a high-powered team of civilian scientists (Drs. Stone, Dutton, Leavitt, and Hall) is summoned to a secret facility to receive the satellite fragments and the survivors from Piedmont. The team expresses some anger, as even the head, Dr. Stone, was unaware of Project Scoop.

Drs. Stone and Hall attempt to stop the spread of the extraterrestrial pathogen by flying over Piedmont and dropping poison canisters intended to kill birds that may have eaten infected flesh from

Chapter 6: Pandemics and Modern Plagues

the dead bodies. Dr. Stone requests the nuclear bombing of Piedmont in hopes of extinguishing the alien agent. The secret facility is itself protected by an automatic nuclear self-destruct mechanism that activates in the event of an accident that releases a pathogen under study at the lab.

The team conducts studies to isolate and characterize the pathogen, which has been named the Andromeda strain, and its transmission. They learn that it is a particle 1-2 microns in diameter, grows when it is exposed to an electron beam, and is spread through the air. The team realizes that the purpose of Project Scoop was to obtain the ultimate biological weapon. With that weapon endangering Americans, they conduct many tests to try to understand why a baby and a drunk are the sole survivors from Piedmont.

Dr. Dutton is exposed to Andromeda accidentally and expects to die. Not long thereafter the isolation chamber's gaskets fail. The team begins to put together the meaning of the events and concludes that Andromeda is "mutating." The mutant forms are not lethal, and Dr. Dutton lives. Dr. Hall observes Dr. Leavitt having an epileptic seizure, which leads him to check the computer read-outs of growth data for which she was responsible. He finds that Andromeda only grows within a narrow pH range, 7.39 to 7.43. This finding explains the survival of the baby and the drunk from Piedmont. Both had blood pH readings outside of that range. Everyone is relieved, for the Andromeda strain in the atmosphere over Piedmont is moving toward the ocean, where they believe the higher pH of the ocean will render it harmless.

Chapter 6: Pandemics and Modern Plagues

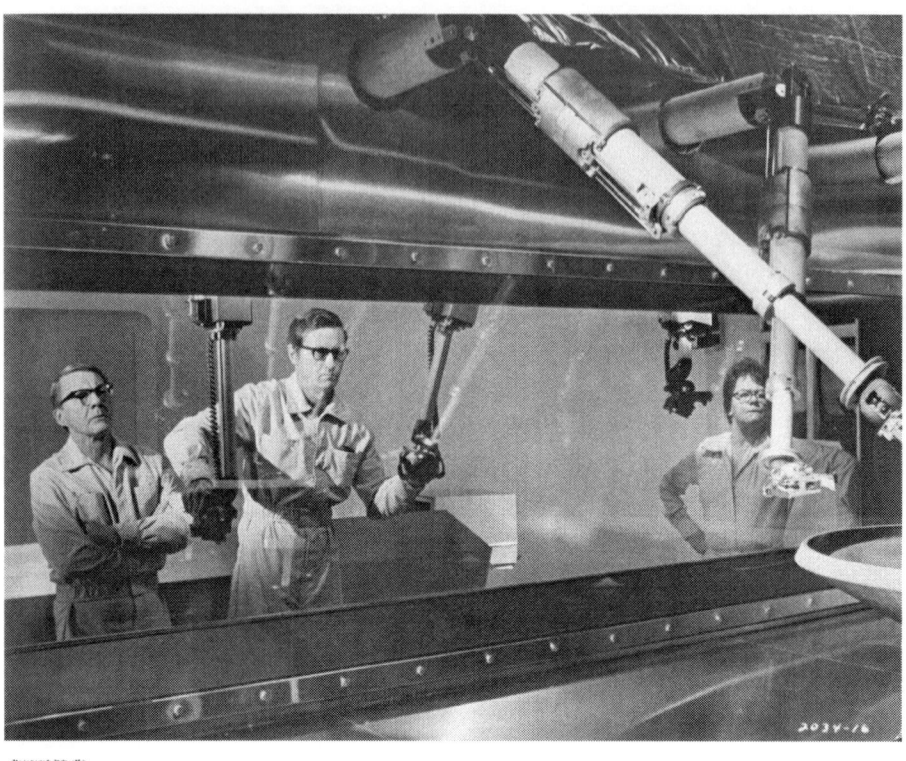

Figure 6. *The Andromeda Strain*: scientists examining a space capsule containing a lethal alien organism, which is inside a sealed containment facility. Courtesy of Universal Pictures/Photofest

A new version of *The Andromeda Strain* was presented on cable television's Arts and Entertainment channel in 2008. The major modification of the storyline was the addition of an investigative reporter. The scientific components were tinkered with significantly, making Andromeda a much less plausible entity. The new version, like the book and the film, succeeds in portraying a deadly biological threat to humanity and explores the entanglement of scientific investigation with national military policy.

Chapter 6: Pandemics and Modern Plagues

SCI-FI SCIENCE vs. REAL SCIENCE

EPIDEMIOLOGY AND PUBLIC HEALTH

OUTBREAK. The dramatic representation of public health professionals trying to solve the path of infection in reverse order to learn the source of the deadly hemorrhagic fever is shown well in this film. We can only hope that in reality there is less impact on the effort by the personal agendas of the professionals and that there is more professionalism exercised. The course and severity of illness in the community is monitored by collecting data on the numbers of cases of illness and death caused by a particular infectious agent.

The death of humans by an agent is called mortality. Mortality is reported in terms of the number of deaths in a given geographic area, such as a city, county, state, or country, or in terms of the number of deaths in a geographic area per some appropriate number of citizens. In a small region that might be as a per cent, which is the number per 100. In a larger region it might be the number of deaths per 10,000 or 100,000. Illness caused by an infectious agent is called morbidity. It is reported similarly to mortality.

The CDC team, upon learning of the outbreak of hemorrhagic fever, discusses the need to identify the carrier and the host. A carrier is a person who harbors an infectious agent, but is not made ill by it. The most famous carrier in American medical history was "Typhoid Mary," who transmitted the disease to many, but, herself, did not fall ill. The team also identifies the need to learn the host for the disease. A host is a non-human species which is infected with the infectious agent, typically without experiencing serious illness or, alternatively, any illness at all. Such a species is also called the reservoir of the infectious agent, because it provides a means for the agent to multiply and be available for transmission to humans.

Chapter 6: Pandemics and Modern Plagues

THE ANDROMEDA STRAIN. There is no doubt that the source of the epidemic that wiped out the small town of Piedmont was the satellite that crashed nearby and was retrieved by the town doctor. One epidemiological problem is the characterization of the agent causing the mortality and design of a plan to prevent spread of the agent. A variety of laboratory tests, both physical and biological, are used to characterize the Andromeda strain and establish its unearthly character. The other aspect is to prevent the spread of the extraterrestrial agent to other locales. The initial action is to kill carrion-eating birds that might transmit Andromeda beyond Piedmont. The destruction of the town with a nuclear bomb is planned to destroy all vestiges of Andromeda on Earth.

Real Science. *Outbreak* is accurate in its representation of the epidemiological goal of tracing the sequence of disease transmission and the source of the first human infection. The process is unlikely to be affected by personal relationships and insubordination in the military as it is in the film. In *The Andromeda Strain* the characterization process is shown in detail and is largely accurate in demonstrating approaches to characterization, which are discussed further below. The efforts to prevent the spread of Andromeda are plausible choices. Both films call attention to the difficulty of stopping an epidemic.

TRANSMISSION OF THE ILLNESS

OUTBREAK. In Zaire Daniels is told not to worry when Major Salt removes his helmet, because the disease is not airborne. He then recounts the beginning of the epidemic which started when an infected man drank from the village well. In several instances the transmission is depicted as occurring via the transfer of bodily fluids. Jimbo gets ill after the monkey spits on him. The pet store owner is scratched by the monkey. Jimbo infects his girlfriend when he kisses her. Robby is pricked with a needle when Casey lurches suddenly because he developed fever-induced convulsions.

Chapter 6: Pandemics and Modern Plagues

In Cedar Creek, a laboratory technician became infected when a tube of the pet store owner's blood broke when he opened the centrifuge. He then went to a movie theater and transmitted the disease by coughing. Casey contracts the disease from the air when his safety suit tears. Daniels figures this out when he is shown an accident victim in the hospital who contracted the mysterious illness without any contact with an infected person. A presidential advisor correctly refers to the virus as airborne in the conference that approves Clean Sweep. The stage is set for a pandemic.

Real Science. Fortunately, the Ebola virus and others like it are not transmitted in the air. They are only transmitted by bodily fluids. Usually this requires an internal fluid such as blood or saliva. It occasionally has been transmitted by skin contact, presumably because sweating leaves virus on the skin. The film cleverly converts the virus to a rapidly transmitted, pandemic-causing agent when Motaba is presented as having mutated so that it survives outside of the body long enough to accomplish airborne transmission. We all must hope that this does not occur in our real world.

THE ANDROMEDA STRAIN. The film shows that the Andromeda strain is transmitted in the air. Dr. Dutton performs a test to determine its size. He uses remotely operated manipulators to insert filters of progressively larger size in the air line to the box containing a test rat. In this way he finds that the Andromeda strain is between 1 and 2 micrometers in diameter. Such a small particle would be distributed through the air readily.

Real Science. The laboratory investigations are good representations of transmission studies for airborne pathogens. Introducing air containing the pathogen is the usual way of checking for transmission by air. Using filters with pores of a precise size is a technique that is utilized in real laboratory investigations.

Because no known pathogens kill as rapidly as the Andromeda strain, there is normally a period of maintaining the test animal during the incubation period until tests for the illness are performed in an appropriate number of days.

BLEEDING BY PEOPLE WITH A MOTABA INFECTION

OUTBREAK. People with the infection have multiple bruises and are bleeding from their eyes, ears, noses, and mouths. Their appearance is terrifying. The film makers put abundant hemorrhage in their depiction of this viral hemorrhagic fever.

Real Science. Many people who are infected with the Ebola virus, whose symptoms are, for all practical purposes, identical to the Motaba virus, do not experience the copious bleeding shown in the film. While such bleeding does occur in some patients, it is more typical for someone to die after a period of feeling listless as organ failure occurs. This is thought to be due to the formation of blood clots in small vessels. The blockages prevent perfusion of the tissues with life-sustaining blood. As a result, regions beyond the blockage die. During an inflammatory response the blood platelets become more apt to initiate a clot, and the liver produces large amounts of fibrinogen, which is the precursor to the major clot protein, fibrin, that forms during blood coagulation.

Those persons who do experience uncontrolled bleeding are thought to have used their clotting factors faster than the liver can manufacture replacements. At one time it was believed that the bleeding was due to the Ebola virus infection in the liver. Cell death and dysfunction were believed to lead to inability on the part of the liver to produce adequate levels of clotting factors for the blood. The resulting bleeding is a failure of <u>hemostasis</u>. The prevention of blood loss by clot formation is called hemostasis. Even though we are unaware of it, the clotting factors are at work around the clock to

Chapter 6: Pandemics and Modern Plagues

prevent bleeding. If it were not for them, we would bleed from the small bangs and bumps of everyday life. (This is the threat faced by hemophiliacs; it is treated by replacing the clotting factor they lack.) When someone has an end-stage Ebola virus infection, bleeding may be evident on the outside of the body, and it occurs internally as well, where it cannot be seen. Internal bleeding can lead to organ failure and death.

TIME COURSE OF THE MOTABA VIRAL ILLNESS

OUTBREAK. In Zaire the local doctor tells Daniels that the disease is not airborne, has a 100% mortality rate, and kills in 2 to 3 days. He says that he does not know the incubation time. Daniels tells General Ford that the incubation time is short. In the U.S. Robby learns during the Boston outbreak that the incubation period is around 24 hours.

Real Science. The time from exposure to the appearance of symptoms of Ebola virus infection may be as long as 21 days or as short as 2 days, but 5 to 10 days is typical. The illness lasts from 7 to 14 days. Mortality ranges from 50% to 90%.

The rapid time to death is a significant factor in the failure of the immune system to defeat a pathogen. It occurs sometimes because some pathogens have a way to defeat the early phase of the immune response.

The human immune system has two divisions, called innate and adaptive immunity. <u>Innate immunity</u> is the first line of defense. It provides generic defense that works against multitudes of pathogens. Immunologists attribute 95% or more of the protection provided by the immune system to innate immunity. There are many ways that innate immunity protects. They can be physical, chemical, or cellular. The skin is an example of physical innate immunity. It provides a barrier

147

which bars access of pathogens to the body interior. The acid of the stomach is an example of chemical immunity. Most microbes that are swallowed are killed or rendered incapable of multiplication in the very acidic environment of the stomach, which secretes hydrochloric acid that also functions in digestion. Cellular innate immunity is provided by cells such as the white blood cells called <u>neutrophils</u> and the tissue cells called <u>macrophages</u>. They are cells which engulf, ingest, and digest particles that they encounter.

<u>Adaptive immunity</u> is initiated by two types of the white blood cells called lymphocytes. When <u>B-lymphocytes</u> recognize antigen, they respond by forming cells that produce <u>antibodies,</u> which are also called <u>immunoglobulins.</u> Antibodies are able to bind antigen and to initiate other defensive responses. One of the most important of these is the activation of <u>complement,</u> a family of proteins that initiates a variety of defensive responses, one of which is inflammation. When <u>T-lymphocytes</u> recognize antigen, they respond with cells called cytotoxic T cells, which can kill target cells (such as a cell infected with a virus) and helper T cells, which activate macrophages to step up their actions beyond what they normally do.

When the human body is under attack by a pathogen for the first time, the innate responses provide the early defense. Many times they succeed in repelling the invader. If an antigen has some special capability that overcomes innate immunity, then adaptive immunity is the next line of defense. Adaptive immunity is very powerful, but it takes at least 5 to 7 days to reach effective levels after the initial recognition of the pathogen. The response reaches peak levels in 10 to 21 days. The adaptive response is rendered useless if the pathogen kills faster than the adaptive response can gear up.

Such is the case with plague and Ebola. Plague succeeds because it attacks macrophages and neutrophils. They are then unable to hold off the bacterium until the adaptive response gears up. Ebola succeeds because it disrupts blood clotting. Organ failure occurs too

Chapter 6: Pandemics and Modern Plagues

fast for the adaptive response to have a chance to save the day.

TREATING THE MOTABA VIRUS DISEASE WITH ANTISERUM

OUTBREAK. Bags of antiserum are administered to people infected with the Motaba virus. The E-1101 antiserum works on the original strain, but not the newer, mutated, airborne strain. Having the host monkey, which Daniels believes is infected with both strains, allows Major Salt to make antiserum for both strains. Patients are given 200 ml intravenously and even those in advanced stages of the disease, including very near death, are saved by the antiserum.

Real Science. An <u>antiserum</u> is a preparation of blood serum, or a portion of blood serum, that contains <u>antibodies</u> or <u>immunoglobulins</u> that have the potential to block the disease processes that are occurring in a patient. (<u>Serum</u> is the liquid that remains after blood clots. <u>Plasma</u> is the liquid part of blood that remains when blood has been prevented from clotting by adding an anti-coagulant and the cells have been removed.) <u>Antibodies</u> are a type of protein produced in response to foreign organisms or substances. They remove or block the action of the inducing material by binding to it, and they can set in motion a variety of responses aimed at controlling and/or eliminating the inducing material. Among these is inflammation. Antibodies form in response to the presence of <u>antigen</u>, foreign material that is recognized by white blood cells called <u>lymphocytes</u>.

The treatment used in *Outbreak* is similar to one that was used during the 1918 influenza pandemic. People who were ill with influenza were given either blood or plasma from people who had recovered from the illness. The treatment was given credit for reducing mortality by as much as 50%. We cannot know whether the success was due to antibodies in the blood of recovered patients or to some other factor. This treatment is one that is being considered in the event of a new influenza pandemic.

The problem with the antiserum solution that saves the day is that it actually takes considerably longer to prepare large scale amounts of antiserum. An antiserum produced at this stage of study of the antigen would most likely be made by injecting the antigen into an animal, which would then respond to the foreign material by forming antibodies to it. They would probably first be detectable in a few days, but the concentration would be too low to be useful as an antiserum. It would be typical to boost the antibody response by injecting the antigen two or three times more. A typical protocol would boost at 21 to 28 days, 42-56 days, and 63-84 days. Using rabbits, which are of a comparable size to a small monkey, the process can be expected to yield 80 - 120 ml of antiserum. The concentration of useful antibody can vary considerably. Often the technique needs to be tinkered with to obtain a strong antibody response. The antiserum solution seems improbable also because there does not appear to be any way to produce so many treatments.

SOURCE OF NEW PATHOGENS

OUTBREAK. The Motaba virus interacts with humans in a deep jungle setting that represents a route of introduction of a pathogen from an area that has hitherto been free from a human presence.

THE ANDROMEDA STRAIN. The deadly pathogen of the film was collected in space by a satellite designed to function as a filter to "scoop up" potential pathogens for use in biological warfare. Apparently, after crashing out of control to Earth, the satellite leaked out something lethal when it was opened.

Chapter 6: Pandemics and Modern Plagues

Real Science. The scenario depicted in *Outbreak* is considered quite plausible by many scientists. The combination of new or more frequent encounter with species in formerly remote habitats and mutation is also plausible. HIV is a candidate for this route. It is known that HIV infects certain monkeys without causing the serious illness it does in humans. If the virus that normally infects the monkeys mutated so that it could recognize human helper T cells, then it could make the transition in host specificity to become the all-too-familiar human pathogen.

As for the prospect of life in outer space that is presented in *The Andromeda Strain*, some scientists believe not only that life can survive in space, but also that the first living things on Earth came here from space. This idea is called <u>panspermia</u>. Some scientists have proposed that the origin of life on Earth was not an evolutionary process that formed cells from non-living materials of the Earth that became more organized and more complex, but instead the first life on Earth came falling through the atmosphere from space, already having diverse, complex molecules. This process, at the very least, would jump-start the evolution of life on Earth. Samples from space have provided little evidence to support this idea, as the molecules found, while more complex than some believed possible, such as the amino acid glycine, are far simpler than the molecules necessary for life. These reports have frequently been contested.

Some believe that very primitive cells could themselves have arrived from space. Whether that is possible, given the presence of the atmosphere, is also a subject of debate. Numerous claims of the recovery of microbes from the upper atmosphere, as high as 41 kilometers, have been made. They have been disputed. One of the more recent and intriguing is the red rain of Kerala, which occurred in 2001. Terrestrial explanations have been proposed, such as a dust cloud containing fungal spores. Dr. Godfrey Louis proposes that the cloud contained primitive cells

Chapter 6: Pandemics and Modern Plagues

lacking DNA but capable of replicating at 300° C. Dr. Chandra Wickramasinghe reports that he did detect DNA in red rain material.

A variety of ideas for the source of microbes from space has been suggested. Hoyle and Wickramasinghe favor comets as the delivery system. The planet Venus has been suggested as a source of microbes detected on photographic plates at the Lockyer Observatory in England. Venus has been suggested as the source of the 1918 influenza pandemic, also.

Studying the possibility of the presence of microbes in space has become more difficult because of human excursions into space. Now, finding a bacterium in space could happen because humans contaminated space with it during a mission, whether manned or unmanned.

A human presence in space also makes possible the introduction of dangerous microbes from space. This seems less improbable now that on Earth microbial life has been found in inhospitable environments that were previously regarded as unable or very unlikely to harbor any living things. The microbes that live in these extreme environments are called <u>extremophiles</u>. Some still believe that such organisms, or at least some of the reported ones, are contaminants. Evidence is mounting that they are not. Extremophiles have been discovered that tolerate high temperature, low temperature, high pressure, high salinity, high levels of metals, high osmotic strength, high pH, low pH, high levels of ionizing radiation, and low levels of moisture. Deep within the earth bacteria have been found that live in the cracks between rocks.

There is now a considerable body of information about extremophiles. One such organism is the bacterium *Methanococcus jannaschii*. It was discovered at a hydrothermal vent, where temperatures are near the boiling point of water, 2,600 meters below sea level in the Pacific Ocean. Thus it displays unusual tolerance of high temperature and high pressure. It is so thoroughly studied that its genome has been sequenced. Consistent with its lifestyle's being very different from ordinary bacteria, it had only a 44% match of its genes to other known genes. In comparison to a very common bacterium, *Hemophilus influenzae*, only 17% matching sequences were found.

CHARACTERISTICS OF NEW PATHOGENS

OUTBREAK. Like the Ebola virus, the Motaba virus is a filovirus. The Motaba virus is very similar to the Ebola virus in its appearance, the symptoms it produces, continent of known origin, and high mortality during outbreaks. These similarities make the biological underpinning of the story quite plausible.

Real Science. As noted above (p. 146), the film overstates the occurrence of visible hemorrhage in people with a Motaba virus infection. There are noteworthy differences between the Motaba virus and the Ebola virus. The time course of the Motaba illness is much faster. It is said to incubate only 24 hours until the onset of symptoms, to kill in 2 to 3 days, and to have a 100% mortality rate. The Ebola virus incubates from 2 to 21 days (5 to 10 days is typical), kills in 7 to 14 days, and has a mortality range of 50-90%. The Motaba virus is transmissible by air; the Ebola virus is transmitted by transfer of bodily fluids. Ebola was discovered in 1976. The first Motaba outbreak was 1967, which predates the appearance of Ebola by 9 years. When the Ebola virus was discovered, the thin, thread-like shape was completely new to science. Since then other members of the group, the filoviruses, have been identified.

THE ANDROMEDA STRAIN. It was the job of the Wildfire team to detect and characterize "the organism" which originated from the crashed satellite and to devise a way to control it. The greenish speck of material they find on the mesh of the satellite's scoop gets larger before their eyes as they examine it microscopically. The team members seem completely willing to regard the specimen which they discovered as "alive." and it is named the Andromeda strain. They find only one characteristic of a living thing, growth, and everything else they find out about it points to a structure that is not cellular and, therefore, not an organism, not a living thing. They characterize it as similar to plastic and determine that it has neither amino acids, which are the building blocks of proteins, nor nucleic acids, which are the repositories of genetic information in cells and viruses. Transmission studies employing filters reveal that particles between one and two micrometers in diameter are quickly lethal when transmitted by air to a healthy animal.

As a pathogen the Andromeda strain causes death very rapidly. People experience a brief period prior to death during which they become psychotic. Then their blood clots throughout the body, becoming a dark red powder.

The Andromeda strain infects animals other than humans. A dog, cat, rats, and monkeys, all mammals, are killed by it. Vultures are not. In this multi-species pathogenicity, it is similar to numerous viruses.

Real Science. The size of the deadly particles is consistent with their being small cells. There is no indication of any cellular character, however. Their chemical make-up is totally unlike the familiar cells of Earth's organisms, for no cell exists that is devoid of amino acids and protein. Even viruses, which are not cells, contain amino acids and nucleic acids.

Viruses bear some similarity to the unknown specimen. They are often considerably smaller than this specimen, but some share with it a crystalline structure, which is revealed in the electron microscope as particles that have geometric shapes. Examination with the electron microscope prompts the growth and reproduction of the strange material, even though it is in a vacuum and being bombarded with a high voltage electron beam. These characteristics are unknown in any organism on Earth, and they are unknown among viruses.

Viruses are very compact structures. They are made up of molecules that pack very efficiently into a particle that is capable of invading a cell and using its systems to reproduce viral molecules, which in turn assemble to form new viral particles. Biologists generally do not regard viruses as organisms because they are not cells and they cannot give rise to new viruses from pre-existing ones without the aid of cells. The Andromeda strain has a crystalline structure like a virus but, unlike a virus, is able to reproduce itself.

If some of the functional characteristics of cells were present in the Andromeda strain, perhaps we would have to conclude that the unearthly thing is functionally a cell even though it lacks the recognized cellular anatomy and metabolism of an Earth cell. The Andromeda strain falls short of that standard.

Studies to detect homeostasis and metabolism were not conducted. The growth studies that are mentioned sound like routine bacterial studies that vary such things as the nutrients that supply energy to the cells. These are difficult to understand in light of the other finding that Andromeda uses high energy sources directly to provide for its energy needs.

Overall, the film has little evidence that the Andromeda strain is a living thing. The fact that the film's scientists do consider it living, however, brings up the intriguing possibility that living things can survive in space. Most scientists believe that extreme conditions of temperature and pressure coupled with exposure to radiation make space an environment in which life cannot survive.

The Andromeda Strain reminds us that the specimens we encounter may not always be easy to place into discrete categories.

Chapter 6: Pandemics and Modern Plagues

The levels of organization found in nature move from molecular assemblages called organelles to functioning groups of organelles called cells. Organelles are not the only molecular assemblages that are found in nature, for viruses constitute a molecular assemblage. Some biologists favor including viruses among living things, because they have concluded that the deciding factor should be, not cellular structure, but the use of nucleic acid as a genetic code.

Viruses do use nucleic acids for their genetic code. They differ from cells in that all cells use deoxyribonucleic acid (DNA) for their repository of genetic information, while viruses, as a group, use either DNA or ribonucleic acid (RNA). This distinction is a way of dividing viruses into two large groups--the DNA viruses and the RNA viruses. The criterion of using a nucleic acid for a genetic code places the emphasis in defining life on the process of reproduction of an identity genetically encoded by a nucleic acid. Viruses do have the ability to produce offspring of their own kind, just as cells do. They are unable to do so without enlisting the services of cells. To some biologists this dependence on cells disqualifies viruses from the roster of living things.

BIOLOGICAL AGENTS FOR WAR OR TERROR

OUTBREAK. The Motaba virus would be a prime candidate for biological warfare or bioterror because, having been kept secret, there is no way for an enemy to treat it. The frozen antiserum that dates to the original outbreak of the Motaba virus in 1967 could be used for normal medical research purposes or defensively should another nation or terrorist group develop and use the virus in biological warfare. The small amount of antiserum limits the scope of its protection.

THE ANDROMEDA STRAIN. Project Scoop is an attempt to scoop up microbes from space and in so doing to scoop the rest of the world in the arena of germ warfare. This strategy succeeds only if the

Chapter 6: Pandemics and Modern Plagues

deadly agent is successfully contained.

Real Science. Real science and politics intersect in this instance. Most aspects of the defensive and offensive activities of the United States are revealed to the public only partially or not at all, just as Project Scoop was a secret activity. In recent years there has been some public examination of the possible need of a program of immunization to protect citizens from biological warfare or bioterrorism. There has also been discussion of whether to stockpile antibiotics or to make certain that large quantities could be supplied in a short time should they be needed. A flurry of such discussion followed the post-9/11 anthrax scare of 2001. Interest in the subject has waned since no new threats have surfaced. The citizens of the U.S. most likely have only fragmentary knowledge of the government's activities and plans in this arena.

There has been public debate about whether nations should keep frozen stocks of smallpox virus, which is considered a possible germ warfare agent. The virus is believed to have been eradicated from the natural world. Consequently, routine immunization for smallpox has ended and created populations vulnerable to the virus via germ warfare. The position of the United States has been to keep stocks for defensive purposes as long as there is reason to believe that other nations, especially potential adversaries, have stocks of the virus.

SAFETY TECHNIQUES & DANGEROUS MICROBES

OUTBREAK. The film depicts working with dangerous microbes very well. The techniques shown are excellent representations of contemporary methods to protect workers from the dangerous microbes they encounter, both in laboratory situations in which they know the identity of the microbe, and in field and laboratory situations in which the microbe's identity is unknown.

Chapter 6: Pandemics and Modern Plagues

Robby and Casey fail to report breaks in technique when they occur. This all-too-human temptation poses danger to fellow workers and sometimes others as well.

THE ANDROMEDA STRAIN. The film shows some techniques that are standard and some that are introduced to heighten the importance of preventing the dissemination of alien microbes. Upon returning from Piedmont, Drs. Stone and Hall are shown to be ridding themselves of contamination from that hot zone. However, the Andromeda strain is later shown to absorb radiation. It seems improbable that the treatment intended to protect would do so.

The Wildfire team members wear protective suits when working with the satellite material or the infected patients. This is an accepted procedure.

During the team's preparations to enter the research area of level 5, they are given immunizations, which are said to be booster shots for the standard diseases. Booster shots are not useful for several days at best. Since the work on the satellite begins the next day, the shots are not helpful. It is not clear why the team is regarded as being in danger of encountering standard diseases.

Because the facility is designed to deal with unknown microbes whose presence is due to germ warfare, it aimed both to protect the scientific team and to make level 5, in which studies of the unknown agent occur, so germ-free that it will be possible to study the agent without fear of contamination from the environment. (See question 19.) They undergo various decontamination techniques. The more stringent ones are fanciful. The surface sterilization process that burns the epithelium to a fine ash is not a technique that is recognized. Despite submitting to the assault on their skin, the team members are not really freed of surface contaminants because they wore protective masks to cover their heads. As a result their facial and neck skin and their scalps are not disinfected.

Chapter 6: Pandemics and Modern Plagues

The team is oblivious to the contamination of Andromeda that occurred when the satellite was opened by the doctor in Piedmont. They really should have found a way to purify Andromeda. The results of the tests are only meaningful if they are performed on an uncontaminated specimen of the space microbe.

Real Science. The films succeed in showing containment and protection techniques for both field work and laboratory studies. The lack of professional integrity displayed by the characters in *Outbreak* who fail to report the breaks in technique is hopefully a behavior that is encountered rarely, if at all, in health care facilities dealing with extremely dangerous pathogens for which no cure is known.

Both films touch on the categories of danger and, thus, the stringency of the techniques used in handling microbes that are associated with biosafety levels. The work area for microbes of the highest biosafety designation, biosafety level 4, is also called a hot zone. Lower levels are cold, neutral, or warm zones (Table 4).

In *The Andromeda Strain* the secret laboratory is presented as a five-level facility. Each level is supposed to use more stringent protective measures, and the final level permits laboratory studies of the most dangerous microbes. Note that these designations do not correspond to the actual biosafety levels (Table 4).

OFFICIAL RESPONSE TO THE OUTBREAK

OUTBREAK. The outbreak in Boston is quelled by sequestering the infected people, and the chain of transmission is broken. Local health officials as well as Robby and her CDC team are responsible for the response. On the west coast, however, the problem is more difficult. An entire town has become infected. When Robby arrives at Cedar Creek, it appears to be under martial law already.

Soon the military, seeing the transmission continuing and fearing the disease will spread beyond the town, decides upon a presidentially approved plan of dropping a bomb on the town to destroy it, and the virus, completely. All inhabitants will be destroyed, whether or not they are infected.

Meanwhile, certain people who were involved in dealing with the original outbreak in 1967, including General McClintock and General Ford, are aware of the existence of frozen antiserum to the virus. They disagree on whether to use the antiserum because it would make evident the secret episode of the first outbreak. For a time the superior officer's decision stands, but eventually, General Ford's conscience gets the better of him. He violates General McClintock's orders, reveals the existence of the antiserum, and allows its use.

Sam figures out that the original Motaba strain has mutated. His actions, which are contrary to orders, may not be an official response. They do come from officials of the federal government who are on the scene to deal with the Motaba crisis. After capturing the monkey that carries the virus, his staff succeeds in preparing antiserum for all in the town. The fictions of this remedy are discussed above (p. 149).

THE ANDROMEDA STRAIN. The official response includes the activation of the Wildfire team of scientists. After studying the Andromeda strain, Dr. Stone asks the president to call a "7-12," a plan to utilize military personnel to destroy Piedmont with a nuclear bomb.

Real Science. In the years between *The Andromeda Strain*, 1970, and *Outbreak*, 1995, the cinematic solution to the prospect of a devastating pandemic does not change: a presidentially approved bombing to destroy all vestiges of the pathogen. With more public awareness of the possibility of this type of disaster, there has also been more questioning about plans to preserve public safety. In the case of

Chapter 6: Pandemics and Modern Plagues

the anthrax scare after the September 11, 2001, terror attacks, the public in the U.S. learned that there were inadequate supplies of antibiotics to treat anthrax. In considering the possibility of attacks with smallpox, the public learned that there were too few doses available for immunization against smallpox. These examples suggest that public health authorities either do not yet take seriously the threat of pandemic illness, whether occurring on purpose or accidentally, or else they choose to keep secret from the public their response plans to protect the public's health.

Chapter 6: Pandemics and Modern Plagues

MOVIE TRIVIA

OUTBREAK

Dustin Hoffman's role of Sam Daniels was originally intended for Harrison Ford. A balance of sorts exists because Harrison Ford's character in *Blade Runner* was originally intended for Dustin Hoffman.

The photos shown as the Motaba virus are actually Ebola. Motaba was said to be in the same family as Ebola, the Filoviridae. As a result it would be expected to resemble Ebola, a long, thread-like virus.

The original end to the film showed the bomb exploding over Cedar Creek. After the film was first screened, the ending was changed, and the town was saved. Presumably this reflects the reception of the more realistic ending.

The monkey with the important role in the story of introducing Motaba into Cedar Creek is a white-headed capuchin. This species is native to Central and South America, not Africa.

The protective suits worn by the soldiers were real, functional suits, not props. Likewise, the masks were real, a type known as M-17. The suit Dustin Hoffman wore is on display in the Planet Hollywood restaurant in the Disney Village section of Disneyland Resort Paris.

The scenes that were supposed to take place in Africa were filmed in Kauai, Hawaii.

The film was expected to have a competitor with a similar story, *Crisis in the Hot Zone*, inspired by the best-selling factual book,

Chapter 6: Pandemics and Modern Plagues

Hot Zone, by Richard Preston. Script problems shelved *Crisis in the Hot Zone*, and *Outbreak* had the infectious disease movie market to itself.

Chapter 6: Pandemics and Modern Plagues

QUESTIONS

1. Do you think it plausible that a bomb would be dropped on an American target to protect the larger population from an uncontrollable, deadly pathogen? What responsibility to the world does a country have when a deadly outbreak occurs within its borders? What role should scientists have in the decision-making process? What role should the Surgeon General have?

2. Weaknesses in the plan to control the Motaba virus and the manner in which it was carried out led to greater loss of life than might have occurred otherwise. Identify weaknesses in the plan and its conduct and suggest improvements.

3. What elements of response to an epidemic are shared in the needed response to a natural disaster, such as Hurricane Katrina, and an attack by an enemy, such as the 9/11 terrorist attack on New York City?

4. How was the Motaba virus introduced into a human in 1967? Is that a plausible explanation? Is it plausible that the virus did not reappear in humans locally for decades? Explain your answer.

5. The proposition that new pathogens are emerging because of human destruction of habitats previously not contacted by humans is controversial. Argue each side.

6. Another explanation for the appearance of new pathogens that has been suggested is that they are the results of research in biological warfare that have breached the containment measures of governments or terrorists doing the research. Present arguments for or against the plausibility of this idea.

7. Do you think an antigenic shift in influenza would be more problematic in going from H1N1 to H1N2 or H2N2? Explain.

Chapter 6: Pandemics and Modern Plagues

8. Ducks infected with H5N1 influenza get a mild illness or no illness at all. Humans get a very serious, often deadly illness. What do you think explains these differences?

9. Does the occurrence of antibiotic resistance qualify as an epidemic or pandemic? Why or why not? If so, characterize it in terms of the WHO phases.

10. Does research and stockpiling of biological warfare agents increase the risk of humankind's suffering a devastating loss of life if it is used intentionally or released accidentally? If so, is the defensive value sufficient to warrant the continued preparation for biological warfare?

11. If a pandemic of monumental proportions were to occur, what do you think the consequences would be?

12. How does the integrity of the health care personnel and the researchers in these two films affect the respective missions? Compare the behaviors of Casey, Robby, and Drs. Dutton and Leavitt.

13. Which diseases are most likely to cause a devastating, global pandemic in your lifetime? Predict the effects of such a pandemic. Is the outcome dependent on which disease occurs? Are there escalating, but less prominent diseases that arguably constitute pandemics occurring now? If so, predict their future impact.

14. Are prion diseases a plausible large scale threat? Consider the current efforts to contain mad cow disease and try to predict the success or failure of those efforts.

15. Are prions, viroids, or spiroplasmas potentially useful as agents of biological warfare or terror?

Chapter 6: Pandemics and Modern Plagues

16. The argument has been made that the more likely way pathogens will affect the human condition is via slow infections or stealth infections. What factors favor this view? What factors argue against it?

17. Should the Surgeon General make adequate protection from known threats for pandemics a health priority in the U.S.? Explain your answer.

18. Do you think a facility like that in *The Andromeda Strain* actually exists? Why or why not?

19. What techniques do the scientists and doctors use to protect themselves from the pathogens? Are there instances where they were in danger, but the film overlooks that? In *The Andromeda Strain* part of the preparation for working with Andromeda is to become as germ-free as possible, including the gut. Is this a useful goal? Would the measures employed achieve it?

20. If birds could transmit the Andromeda strain, do you think the control measure employed would have stopped it from occurring? Why do you say so?

21. When Stone exposes a rat to the air in a room with the satellite, it dies. If the rat had lived, would that have confirmed that there was no longer a pathogen associated with the satellite?

22. When Hall and Anson work with their patients, the baby and the old man, in which room are they located? Explain.

23. Do you think humanity is at a greater risk from naturally occurring biological agents or special strains stockpiled by nations for biological warfare? Explain your answer.

Chapter 6: Pandemics and Modern Plagues

24. If you have access to the 2008 production of *The Andromeda Strain*, compare Andromeda in the older film version with the newer TV version. Which characteristics of the newer Andromeda seem farfetched in light of current science?

25. If you have access to the 2008 production of *The Andromeda Strain*, identify variations in the scientific investigation of Andromeda, compare them with the film version, and assess the extent to which they are plausible, realistic, and useful.

Chapter 6: Pandemics and Modern Plagues

Table 1
Phases of the Influenza Pandemic Cycle

Interpandemic Period	
Phase 1	No new influenza virus subtypes found in humans
Phase 2	No new subtypes, but animal forms pose a substantial risk
Pandemic Alert Period	
Phase 3	Human infections with new subtype, no human-to-human transmission
Phase 4	Small clusters, limited human-to-human spread
Phase 5	Larger clusters, localized human-to-human spread
Pandemic Period	
Phase 6	Pandemic, increased and sustained transmission in general population

Table 2
Some Strains of Influenza A*

Human Strain	Antigenic Subtype
Puerto Rico/8/34	H0N1
Fort Monmouth/1/47	H1N1
Singapore/1/57	H2N2
Hong Kong/1/68	H3N2
USSR/80/77	H1N1
Brazil/11/78	H1N1
Bangkok/1/79	H3N2
Taiwan/1/86	H1N1

Non-human strains	Antigenic Subtype
Sw/Iowa/15/30**	H1N1
Sw/Taiwan/70	H3N2
Eq/Prague/1/56**	H7N7
Eq/Miami/1/63	H3N8
Fowl/Dutch/27	H7N7
Tern/South America/61	H5N3
Turkey/Ontario/68	H8N4

*Strain names indicate the animal host, the geographic location of the infected animal, the laboratory isolate number, and the year isolated.

**Sw = swine
  Eq = equine

Chapter 6: Pandemics and Modern Plagues

Table 3

Examples of Bioterror Agents

Category A	Anthrax, smallpox, botulinum toxin, Ebola, plague, Marburg, tularemia
Category B	Brucellosis, psittacosis, Q fever, ricin toxin, staphylococcal enterotoxin B, typhus, viral encephalitis, cholera (water supply)
Category C	Hanta virus, multi-drug-resistant tuberculosis, nipah virus

Chapter 6: Pandemics and Modern Plagues

Table 4

Biosafety Levels for Safely Working with Microbes

Biosafety Level 1 (BSL1)	microbes not known to cause disease in healthy humans	cold zone
Biosafety Level 2 (BSL2)	microbes with moderate hazard potential, some may be pathogens	cold zone
Biosafety Level 3 (BSL3)	pathogens which cause serious, sometimes lethal, disease	neutral or warm zone
Biosafety Level 4 (BSL4)	pathogens very likely to kill	hot zone